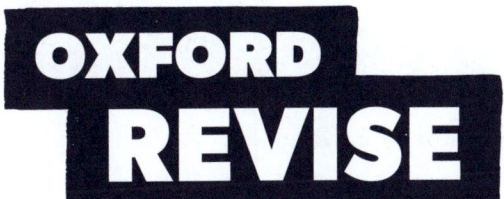

EDEXCEL GCSE
MATHS
Higher

COMPLETE REVISION AND PRACTICE

Naomi Bartholomew-Millar
Paul Hunt
Victoria Trumper

Contents

 Shade in each level of the circle as you feel more confident and ready for your exam.

How to use this book iv

1 Calculating with all four operations, place value, powers and indices 2
- Knowledge
- Retrieval
- Practice

2 Rounding, truncating, error intervals, and estimating 10
- Knowledge
- Retrieval
- Practice

3 Factors, multiples, primes, standard form, and surds 16
- Knowledge
- Retrieval
- Practice

4 Fractions, decimals, percentages 26
- Knowledge
- Retrieval
- Practice

5 Algebra basics 34
- Knowledge
- Retrieval
- Practice

6 Linear graphs 42
- Knowledge
- Retrieval
- Practice

7 Real-life graphs 50
- Knowledge
- Retrieval
- Practice

8 Solving inequalities in 1 or 2 variables 57
- Knowledge
- Retrieval
- Practice

9 Factorising quadratic expressions and solving quadratic equations 62
- Knowledge
- Retrieval
- Practice

10 Quadratic graphs, iterations, solving quadratic inequalities 70
- Knowledge
- Retrieval
- Practice

11 Solving simultaneous equations 78
- Knowledge
- Retrieval
- Practice

12 Sequences 86
- Knowledge
- Retrieval
- Practice

13 Cubic graphs, reciprocal graphs, exponential graphs, transformation of graphs 94
- Knowledge
- Retrieval
- Practice

14 Non-linear real-life graphs 102
- Knowledge
- Retrieval
- Practice

15 Algebraic fractions, rearranging formulae with algebraic fractions, proof, functions and composite functions **108**
- ⚙ Knowledge ⊖
- ↔ Retrieval ⊖
- ✎ Practice ⊖

16 Ratio **118**
- ⚙ Knowledge ⊖
- ↔ Retrieval ⊖
- ✎ Practice ⊖

17 Compound measures and multiplicative reasoning **126**
- ⚙ Knowledge ⊖
- ↔ Retrieval ⊖
- ✎ Practice ⊖

18 Polygons, angles, and parallel lines **134**
- ⚙ Knowledge ⊖
- ↔ Retrieval ⊖
- ✎ Practice ⊖

19 Area and perimeter (including circles) **143**
- ⚙ Knowledge ⊖
- ↔ Retrieval ⊖
- ✎ Practice ⊖

20 Surface area and volume **152**
- ⚙ Knowledge ⊖
- ↔ Retrieval ⊖
- ✎ Practice ⊖

21 Pythagoras and 2D trigonometry **161**
- ⚙ Knowledge ⊖
- ↔ Retrieval ⊖
- ✎ Practice ⊖

22 Similarity and congruence **171**
- ⚙ Knowledge ⊖
- ↔ Retrieval ⊖
- ✎ Practice ⊖

23 Transformations **180**
- ⚙ Knowledge ⊖
- ↔ Retrieval ⊖
- ✎ Practice ⊖

24 Plans, elevations, constructions, bearings **187**
- ⚙ Knowledge ⊖
- ↔ Retrieval ⊖
- ✎ Practice ⊖

25 Trigonometry in 3D, sine and cosine rules **196**
- ⚙ Knowledge ⊖
- ↔ Retrieval ⊖
- ✎ Practice ⊖

26 Circle theorems and circle geometry **204**
- ⚙ Knowledge ⊖
- ↔ Retrieval ⊖
- ✎ Practice ⊖

27 Vectors **212**
- ⚙ Knowledge ⊖
- ↔ Retrieval ⊖
- ✎ Practice ⊖

28 Probability **218**
- ⚙ Knowledge ⊖
- ↔ Retrieval ⊖
- ✎ Practice ⊖

29 Tables, averages, and range **228**
- ⚙ Knowledge ⊖
- ↔ Retrieval ⊖
- ✎ Practice ⊖

30 Charts and graphs **234**
- ⚙ Knowledge ⊖
- ↔ Retrieval ⊖
- ✎ Practice ⊖

31 Data collection, cumulative frequency, box plots **242**
- ⚙ Knowledge ⊖
- ↔ Retrieval ⊖
- ✎ Practice ⊖

How to use this book

This book uses a three-step approach to revision: **Knowledge**, **Retrieval**, and **Practice**.
It is important that you do all three; they work together to make your revision effective.

Knowledge

Knowledge comes first. Each chapter starts with a **Knowledge Organiser**. These are clear easy-to-understand, concise summaries of the content that you need to know for your exam. The information is organised to show how one idea flows into the next so you can learn how everything is tied together, rather than lots of disconnected facts.

Worked example

Worked examples offer step by step guidance on working through a question, to a solution.

LINK

The **Link** box highlights a reference to a related topic you may want to refer to.

REVISION TIP

Revision tips offer you helpful advice and guidance to aid your revision and help you to understand key concepts and remember them.

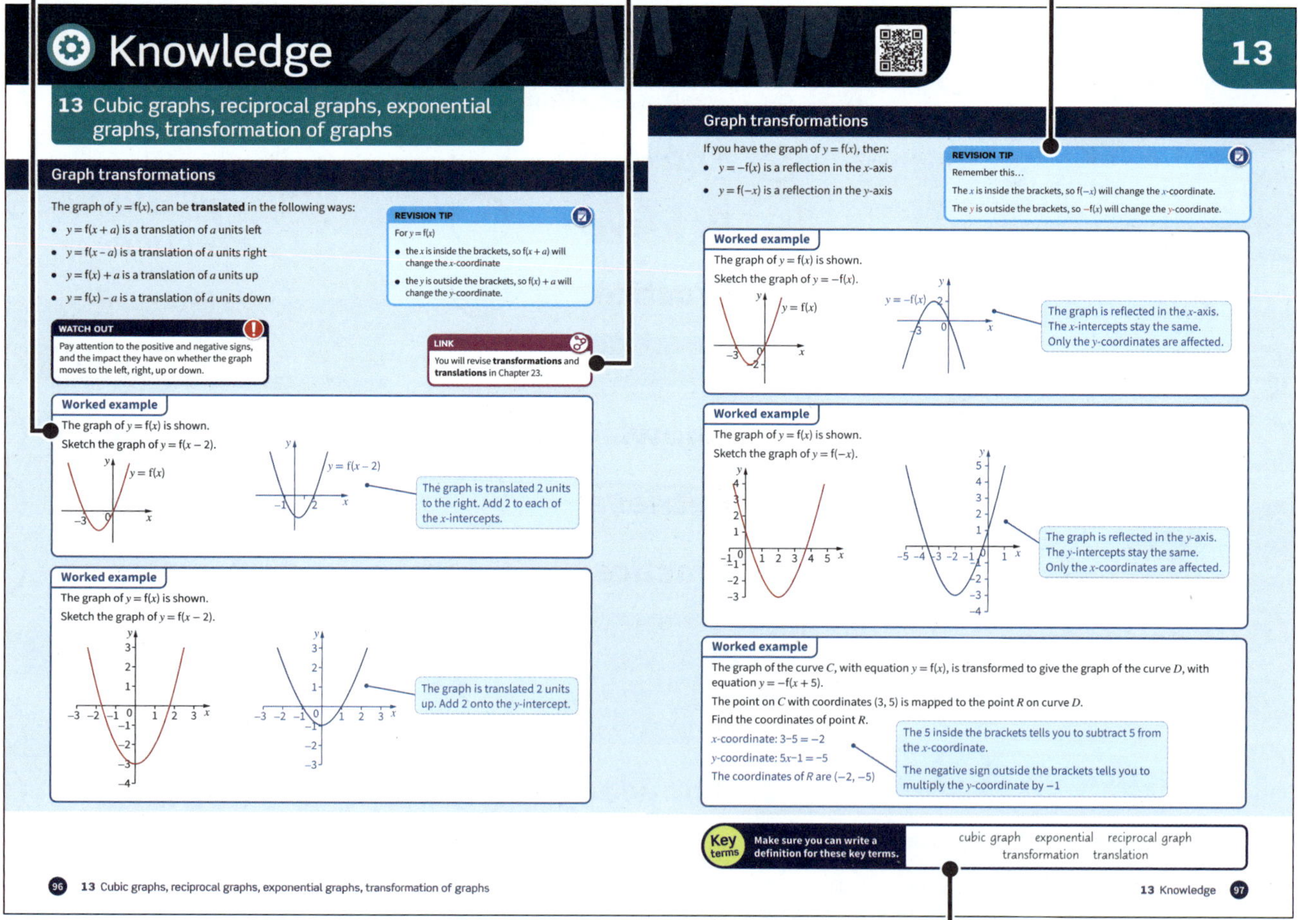

Other features:

Formula

The **Formula** box highlights key formulas you will need to remember.

Key terms — Make sure you can write a definition for these key terms

The **Key terms** box highlights the key words and phrases you need to know, remember and be able to use confidently.

Retrieval

The **Retrieval questions** help you learn and quickly recall the information you've acquired. These are short questions and answers about the content in the Knowledge Organiser you have just reviewed. Cover up the answers with some paper and write down as many answers as you can from memory. Check back to the Knowledge Organiser for any you got wrong, then cover the answers and attempt all the questions again until you can answer *all* the questions correctly.

Make sure you revisit the retrieval questions on different days to help them stick in your memory. You need to write down the answers each time, or say them out loud, otherwise it won't work.

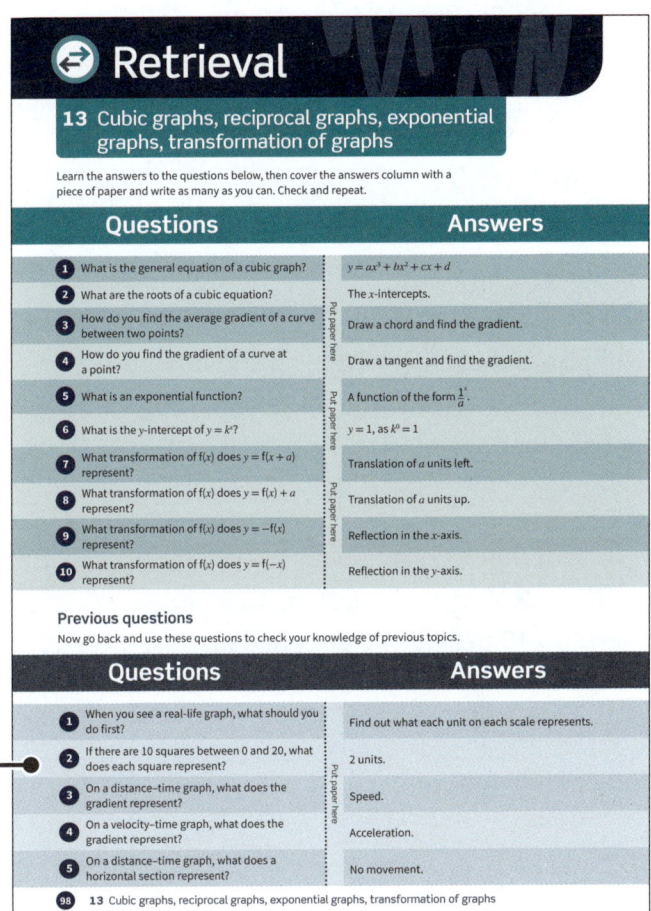

Previous Questions

Each chapter also has some **Retrieval questions** from **previous chapters**. Answer these to see if you can remember the content from the earlier chapters. If you get the answers wrong, go back and do the Retrieval questions for the earlier chapters again.

Practice

Once you think you know the Knowledge Organiser and Retrieval answers really well, you can move on to the final stage: **Practice**.

Each chapter has **exam-style questions**, including some questions from previous chapters, to help you apply all the knowledge you have learnt and can retrieve.

Answers and Glossary

You can scan the QR code at any time to access the sample answers and mark schemes for all the exam-style questions, glossary containing definitions of the key terms, as well as further revision support. Visit go.oup.com/OR/GCSE/Ed/Maths/H

EXAM TIP

Exam tips show you how to interpret the questions, provide guidance on how to answer them, and advice on how to secure as many marks as possible.

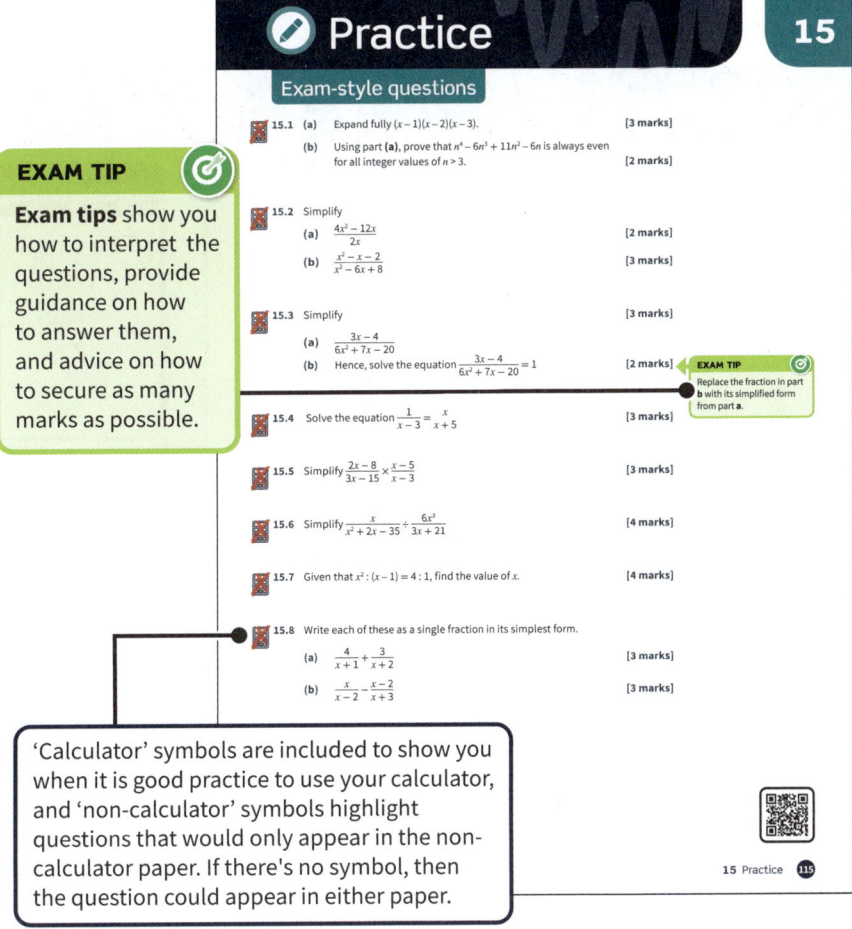

'Calculator' symbols are included to show you when it is good practice to use your calculator, and 'non-calculator' symbols highlight questions that would only appear in the non-calculator paper. If there's no symbol, then the question could appear in either paper.

Knowledge

1 Calculating with all four operations, place value, powers and indices

Definitions

An **integer** is a whole number, such as 1, 73, 946, −17.

The **product** is the result when you multiply numbers.

The **sum** is the result when you add numbers.

The **quotient** is the result when you divide one number by another.

Commutative means that the calculation can be done in any order. For example, addition and multiplication are commutative:

$7 + 4 = 4 + 7$ $8 \times 2 = 2 \times 8$

But subtraction and division are not:

$7 - 4 \neq 4 - 7$ $8 \div 2 \neq 2 \div 8$

Order of operations (BIDMAS)

Use BIDMAS to help you do calculations in the correct order.

Brackets ( )
Indices (powers and roots)
Division and **M**ultplication
Addition and **S**ubtraction

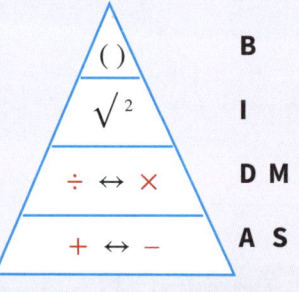

Worked example

1. $43 - 9 \times 4 + 5$

   $43 - \underline{9 \times 4} + 5$ — BIDMAS Multiplication first.
   $= \underline{43 - 36} + 5$ — BIDMAS Addition and Subtraction.
   $= 7 + 5 = 12$

2. $32 - 6^2 \div 2$

   $32 - \underline{6^2} \div 2$ — BIDMAS Indices (powers) first.
   $= 32 - \underline{36 \div 2}$ — BIDMAS Division.
   $= 32 - 18 = 14$

Ordering numbers

Worked example

Which is larger, 1.3 or 1.105?

```
1 . 3 0 0
1 . 1 0 5
```

$1.3 > 1.105$

Write in columns, then write in 0s so both numbers have the same number of decimal places.

Inequality symbols

The small end points to the small number.

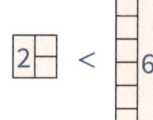 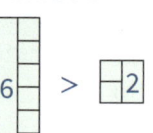

Negative numbers

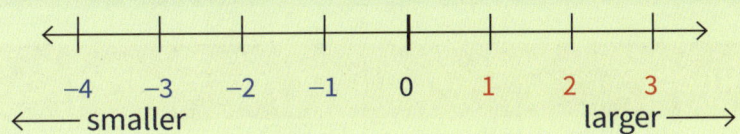

Rules for adding and subtracting negative numbers

Adding a positive number.	Adding a negative number is the same as subtracting.
5 + 2	5 + −2

5 + −2
replace + − with −
= 5 − 2 = 3

Subtracting a positive number.	Subtracting a negative number is the same as adding.
−4 − 3	−4 − −3

−4 − −3
replace − − with +
−4 + 3 = −1

Worked example

1 − 3 − − 6

1 − 3 − − 6   Replace − − with +

= 1 − 3 + 6

= −2 + 6

= 4

Rules for multiplying and dividing negative numbers

positive × or ÷ positive = positive
positive × or ÷ negative = negative
negative × or ÷ positive = negative
negative × or ÷ negative = positive

Worked example

1. 4 × (−2) = −8   positive × negative = negative
2. (−3) × 5 = −15   negative × positive = negative
3. (−3) × (−4) = 12   negative × negative = positive
4. 6 ÷ (−2) = −3   positive ÷ negative = negative
5. (−8) ÷ (−2) = 4   negative ÷ negative = positive

Calculating with decimals

When adding or subtracting decimals:

- Align the decimal points and carry out the calculation as usual
- Make sure to include the decimal point in your answer.

When multiplying or dividing decimals:

- Ignore the decimal points and carry out the calculation as usual
- Then work out where to place the decimal points.

Knowledge

1 Calculating with all four operations, place value, powers and indices

Calculating with decimals

Worked example

Calculate:

1. 21.47 + 32.85

```
  2 1 . 4 7
+ 3 2 . 8 5
-----------
  5 4 . 3 2
      1 1
```

- Align decimal points.
- Add numbers vertically. Start with the right-hand column, and move left.
- Make sure you carry the tenths correctly.

2. 35.7 − 26.34

```
  ²3⁻¹5 . ⁶7⁻¹0
−   2 6 . 3 4
---------------
    9 . 3 6
```

- Align decimal points.
- Add a zero so you have the same number of place values in each number.
- Subtract the second row from the top row. Start with the right-hand column, and move left.
- Make a place value exchange when you need to.

3. 8.16 × 4.9

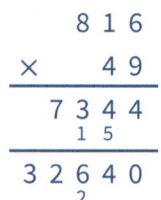

```
      8 1 6
  ×     4 9
-----------
    7 3 4 4
      1 5
    3 2 6 4 0
        2
-----------
    3 9 9 8 4
```

8.16 = 816 ÷ 100
4.9 = 49 ÷ 10
39 984 ÷ 100 ÷ 10
= 39.984

- Ignore decimal points.
- Multiply each digit of the top number by '9'.
- Multiply each digit of the top number by '40'.
- Add the products of the two multiplications together.
- Work out what power of 10 to divide by to find the answer to the original question.

4. 11.07 ÷ 0.9

$$\frac{11.07}{0.9} \xrightarrow{\times 10}_{\times 10} \frac{110.7}{9}$$

```
       1 2 . 3
   9 ) 1 1²0 .²7
```

- Write the division as a fraction and find an equivalent with an integer denominator.
- Carry out the new division working from left to right.

Bases and indices

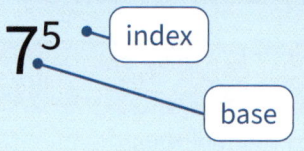

7^5 — index, base

$a^0 = 1$

Example:
$7^0 = 1$

Any number (or letter) raised to the power of 0 is 1.

$a^n \times a^m = a^{n+m}$

Example:
$3^2 \times 3^4 = 3^{2+4} = 3^6$

To multiply numbers with the same base, add the powers.

Bases and indices

$a^n \div a^m = a^{n-m}$	$(a^n)^m = a^{n \times m}$	$a^{-1} = \dfrac{1}{a}$	$a^{\frac{1}{n}} = \sqrt[n]{a}$
Example: $5^7 \div 5^3 = 5^{7-3} = 5^4$	Example: $(4^3)^5 = 4^{3 \times 5} = 4^{15}$	Example: $2^{-1} = \dfrac{1}{2}$	Example: $8^{\frac{1}{3}} = \sqrt[3]{8}$
To divide numbers with the same base, subtract the powers.	To raise a power by another power, multiply the powers.	A number raised to the power of –1 gives the reciprocal of the number.	A power of $\dfrac{1}{n}$ gives the nth root.

An indices question will either ask you to calculate or simplify.

- **Calculate** (or work out) means find the numerical answer.
- **Simplify** means write in a simpler form, but your answer will still contain powers.

Worked example

Calculate:

(a) $\left(\dfrac{9}{4}\right)^{\frac{1}{2}}$

$\left(\dfrac{9}{4}\right)^{\frac{1}{2}} = \sqrt{\dfrac{9}{4}}$

$= \dfrac{\sqrt{9}}{\sqrt{4}}$

$= \dfrac{3}{2}$

Use $a^{\frac{1}{n}} = \sqrt[n]{a}$

Apply the power to the numerator and denominator separately.

(b) $64^{-\frac{2}{3}}$

$64^{-\frac{2}{3}} = \dfrac{1}{64^{\frac{2}{3}}}$

$= \dfrac{1}{\left(64^{\frac{1}{3}}\right)^2}$

$= \dfrac{1}{\left(\sqrt[3]{64}\right)^2}$

$= \dfrac{1}{(4)^2}$

$= \dfrac{1}{16}$

Use $a^{-1} = \dfrac{1}{a}$

Use $(a^n)^m = a^{n \times m}$

Use $a^{\frac{1}{n}} = \sqrt[n]{a}$

(c) Express $\sqrt{\dfrac{5^3}{5^{-5}}}$ in the form 5^n. State the value of n.

$\sqrt{\dfrac{5^3}{5^{-5}}} = \left(\dfrac{5^3}{5^{-5}}\right)^{\frac{1}{2}}$

$= (5^{3-(-5)})^{\frac{1}{2}}$

$= (5^8)^{\frac{1}{2}}$

$= 5^{8 \times \frac{1}{2}}$

$= 5^4$

Hence, $n = 4$

Use $a^{\frac{1}{n}} = \sqrt[n]{a}$

Use $a^n \div a^m = a^{n-m}$

$(a^n)^m = a^{n \times m}$

Indices on your calculator

The symbols on different calculators vary slightly, but often look like this:

Cube root: Cube: Any root:

Square root: Square: Any power:

1 Knowledge

Knowledge

1 Calculating with all four operations, place value, powers and indices

Listing outcomes

A list or table showing all possible outcomes of an experiment or of **combined experiments** is called the **possibility space** or **sample space**. For a combined experiment:

number of possible outcomes $= m \times n$

Where

- m = possible outcomes of one experiment
- n = possible outcomes of the other experiment

Worked example

A padlock has three dials with digits 0–9 on each dial. How many possible combinations are there?

number of possible combinations =

number of outcomes on dial 1 multiplied by number of outcomes on dial 2 multiplied by number of outcomes on dial 3.

There are 10 possible outcomes for each of the three dials.

Total number of combinations $= 10 \times 10 \times 10$
$= 1000$

Worked example

A football team of 11 players has four attacking players and seven defensive players. Two players are to be selected to receive two different awards on behalf of the team. Two possible options for doing this are given.

Option 1: Select one attacking and one defensive player.

Option 2: Select any two different players.

Find the number of possible combinations for each option.

number of possible outcomes $= m \times n$

Option 1: $4 \times 7 = 28$ combinations

Option 2: $11 \times 10 = 110$ combinations

WATCH OUT

You can't pick the same player twice, so in option 2, you need to subtract the player that gets chosen in the first event from the second event.

Key terms — Make sure you can write a definition for these key terms.

calculate   commutative   combined experiment   cube
possibility space   power   product   quotient
sample   space   simplify   sum

1 Calculating with all four operations, place value, powers and indices

Retrieval 1

1 Calculating with all four operations, place value, powers and indices

Learn the answers to the questions below, then cover the answers column with a piece of paper and write as many as you can. Check and repeat.

	Questions	Answers
1	What is an integer?	A whole number.
2	What does commutative mean?	Can be done in any order.
3	What do the letters of BIDMAS stand for?	Brackets, indices, division, multiplication, addition and subtraction.
4	What does BIDMAS tell you?	The order in which to do a calculation.
5	When you multiply a negative number and a positive number, what sign does the answer have?	Negative.
6	When you divide a negative number by a negative number, what sign does the answer have?	Positive.
7	What is the product of two or more numbers?	The result you get when you multiply them together.
8	What is the sum of two or more numbers?	The result you get when you add them together.
9	What is the quotient of two numbers?	The result you get when you divide one of the numbers by the other.
10	What is a list or table showing all possible outcomes of an experiment called?	A possibility space or sample space.
11	What is the value of any number or variable raised to the power of 0?	1
12	Write examples to show each of the six laws of indices.	$5^0 = 1$ $5^3 \times 5^5 = 5^{3+5} = 5^8$ $5^8 \div 5^5 = 5^{8-5} = 5^3$ $(5^5)^3 = 5^{5 \times 3} = 5^{15}$ $5^{-1} = \dfrac{1}{5}$ $5^{\frac{1}{2}} = \sqrt{5}$ (any base number can be used)
13	Which buttons on your calculator would you use to find 4^6?	$4\;\boxed{x^n}\;6 =$
14	What is $27^{\frac{2}{3}}$ the same as calculating?	$(\sqrt[3]{27})^2$
15	Raising a number to which power gives you the reciprocal of a number?	-1

Practice

Exam-style questions

1.1 Work out:

(a) $25.043 - 17.82$ [2 marks]

(b) $17.12 \div 0.8$ [2 marks]

1.2 Work out:

(a) $\dfrac{4 \times 5^2}{4 \times 5 \div 2}$ [1 mark]

(b) $(1 - 0.1) \times 4 - (-10)$ [2 marks]

(c) $\dfrac{(-0.2) \times (-6)}{-1 + 0.7}$ [2 marks]

1.3 Marina's fence measures 1.4 m by 10.5 m. It costs £0.60 per square metre to paint the fence. How much does it cost to paint the fence in total? [3 marks]

1.4 Thema says that multiplying by 0.01 is the same as dividing by 100. Is Thema correct? Explain your reasoning. [1 mark]

1.5 Peter says that $2^3 \times 5^2$ simplifies to 10^5. Peter is wrong. Explain why. [1 mark]

1.6 Simplify fully $\dfrac{(2^7) \times (2^4)^{-1}}{2}$.

Write your answer as a power of 2 [2 marks]

EXAM TIP
An index of -1 is a reciprocal. An index of $\dfrac{1}{n}$ is a root.

1.7 Simplify fully:

(a) $\left(3^{\frac{1}{4}}\right)^{\frac{1}{4}}$ [1 mark]

(b) $\dfrac{a^{\frac{2}{3}}}{\sqrt{a}}$ [1 mark]

1.8 Work out the value of:

(a) $8^{\frac{2}{3}}$ [2 marks]

(b) $\left(\dfrac{16}{9}\right)^{-\frac{2}{4}}$ [3 marks]

1.9 $3 \times \sqrt{27} = 3^n$

Find the value of n. [3 marks]

1.10 Find the value of the reciprocal of $2\frac{2}{3}$

Write your answer as a decimal. [2 marks]

1.11 $\frac{1}{a} = 2^{-3}$ $2^b = 2\sqrt{2}$ $\frac{1}{\sqrt[c]{2}} = 2^{-\frac{1}{3}}$

Work out the value of $\frac{ab}{c}$. [4 marks]

1.12 Write the following numbers in ascending order. [2 marks]

$\left(-\frac{1}{3}\right)^3, 3^{-2}, \left(-\frac{1}{2}\right)^4, \left(-\frac{1}{4}\right)^{-2}, 2^{-1}$

1.13 A four-digit pin code can only use the digits 1 to 9. Digits can be used more than once.

(a) How many possible pin codes are there? [2 marks]

(b) How many of the pin codes use the digit 8 at least once? [2 marks]

1.14 A shop sells:

15 types of chocolate bar

6 flavours of crisps

8 choices of canned drink.

Isabella wants to choose a chocolate bar and a canned drink **or** a packet of crisps and a canned drink **or** a chocolate bar, a packet of crisps and a canned drink.

Show that there are 888 different combinations for Isabella to choose from. [4 marks]

1.15 In a games competition, there are eight teams.
Each team plays four games against each of the other teams.

Work out how many games are played in total. [2 marks]

Knowledge

2 Rounding, truncating, error intervals, and estimating

Rounding

When numbers are **rounded** they are written only to a given **place value**.

Worked example

Round 15 726 to the nearest 1000.

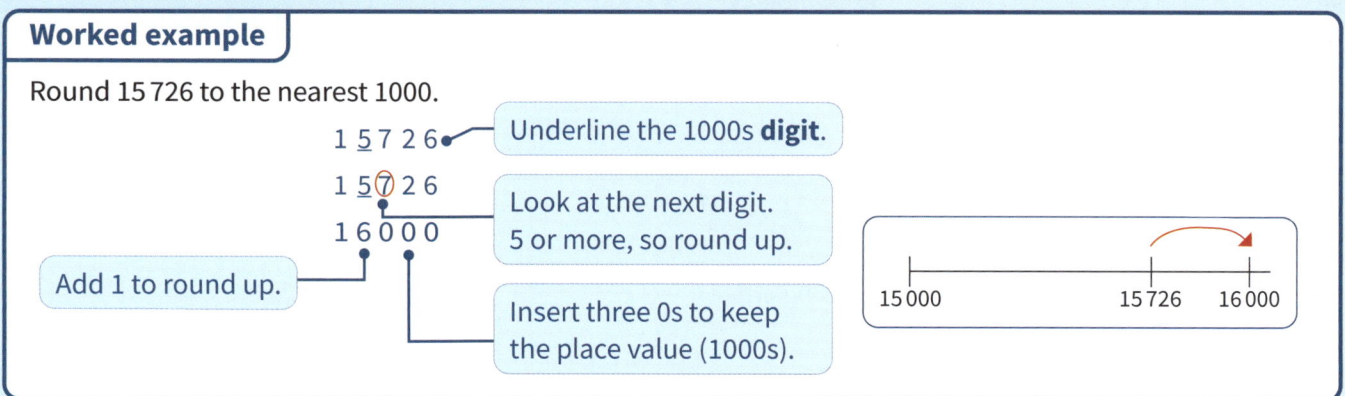

1 5 7 2 6 — Underline the 1000s **digit**.

1 5 ⑦ 2 6 — Look at the next digit. 5 or more, so round up.

1 6 0 0 0

Add 1 to round up.

Insert three 0s to keep the place value (1000s).

Worked example

Round 0.143 76 to 2 decimal places.

0 . 1 4 3 7 6 — Underline the digit in the 2nd decimal place (100ths digit).

0 . 1 4 ③ 7 6 — Look at the next digit. Less than 5, so round down.

0 . 1 4 (2 d.p.) — State the level of accuracy.

Delete the digits after the 2nd decimal place. Don't insert any 0s, as 0.14 has the correct place value (100ths).

Significant figures (s.f.) are digits in a number that contribute to its accuracy.

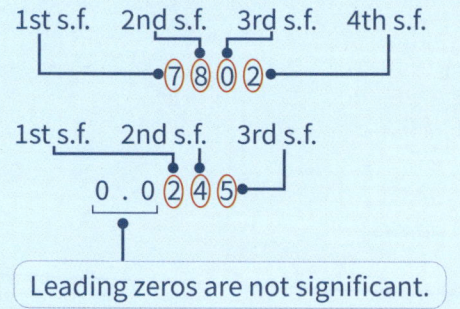

1st s.f. 2nd s.f. 3rd s.f. 4th s.f.
   7   8   0   2

1st s.f. 2nd s.f. 3rd s.f.
0 . 0 ② ④ ⑤

Leading zeros are not significant.

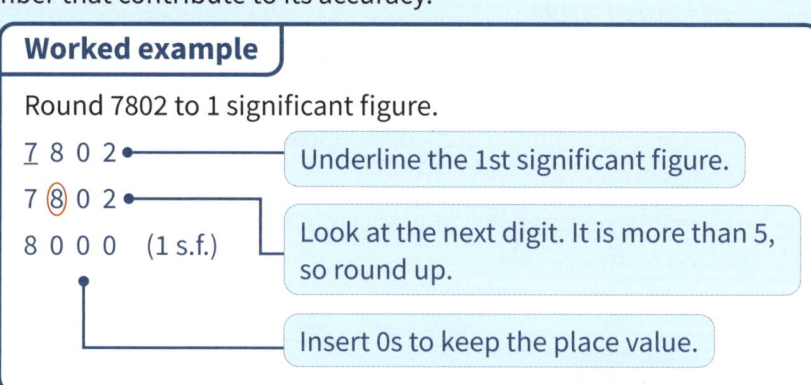

Worked example

Round 7802 to 1 significant figure.

7 8 0 2 — Underline the 1st significant figure.

7 ⑧ 0 2 — Look at the next digit. It is more than 5, so round up.

8 0 0 0 (1 s.f.)

Insert 0s to keep the place value.

Error intervals are inequalities that show possible values of numbers before rounding.

Truncate means removing digits without rounding. 3.7546 truncated to 1 decimal place is 3.7

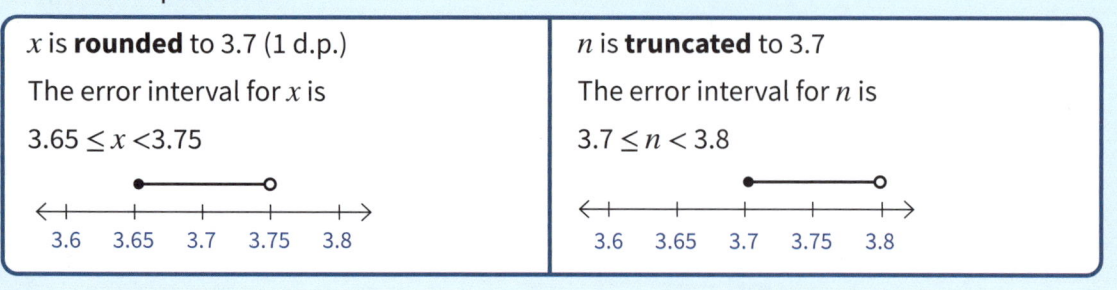

x is **rounded** to 3.7 (1 d.p.)	n is **truncated** to 3.7
The error interval for x is	The error interval for n is
$3.65 \leq x < 3.75$	$3.7 \leq n < 3.8$

LINK

See inequalities in Chapter 8.

Error intervals

Error intervals use inequalities to show possible values of numbers that have been truncated or rounded.

When a number has been **truncated**, the true value can be up to one unit more than the given measurement.

When a number has been **rounded**, the true value could be half a unit less or up to half a unit more than the stated value.

You can use a **number line** to represent the values that are included in the error interval.

Worked example

1. x is **truncated** to 3.7

   Draw the error interval for x on a number line.

   > Minimum is anything that starts with '3.7'
   >
   > Maximum must be below 3.8,
   > So, error interval is $3.7 \leq x < 3.8$

2. x is **rounded** to 1 decimal place.

   The result is 3.7

   Draw the error interval for x on a number line.

   > Minimum could be half a unit less so $x \leq 3.65$
   >
   > Maximum can be up to half a unit more, so $x < 3.75$
   >
   > So, the error interval is $3.65 \leq x < 3.75$

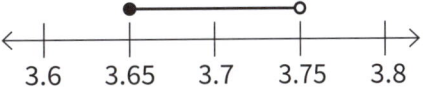

> Draw your solution on a number line.
>
> A hollow circle shows that the number is <u>not</u> included (less/greater than).
>
> A filled-in circle shows that the number <u>is</u> included (less/greater than or equal to).

3. A greetings card has side lengths 11 cm to the nearest centimetre.

   Write the error interval showing the maximum and minimum possible lengths of the sides.

   $10.5 \leq x < 11.5$

   > Anything 10.5 cm or above will be rounded up to 11 cm; anything less than 11.5 cm will be rounded down to 11 cm.

4. The mass of an apple, a g, is given as 80 g. Give the error interval for a if the mass was rounded to the nearest 10 g

   $75 \leq a < 85$

   > Mass was rounded to 1 s.f., so error interval will have 2 s.f., and can be greater than or equal to 5 below, and less than 5 above.

Key terms — Make sure you can write a definition for these key terms.

decimal place   error interval   estimate
number line   place value   rounded
significant figure   truncating

2 Knowledge

Knowledge

2 Rounding, truncating, error intervals, and estimating

Error intervals

Worked example

1. Consider the formula $a = \frac{F}{m}$.

   $F = 45$ to 2 s.f.
   $m = 16.1$ to 3 s.f.

   By considering bounds, work out the value of a to a suitable degree of accuracy.

   $44.5 \leq F < 45.5$ — Find the error intervals for F and m.
   $16.05 \leq m < 16.15$

   lower bound for $a =$
   $\frac{\text{lower bound of } F}{\text{upper bound of } m} = \frac{44.5}{16.15}$ — Find the upper and lower bounds for a.
   $= 2.755...$

   upper bound for $a =$
   $\frac{\text{upper bound of } F}{\text{lower bound of } m} = \frac{45.5}{16.05}$
   $= 2.834...$

   $a = 2.8$ to 2 s.f. — Both the upper and lower bound of a round to 2.8

   REVISION TIP

   In the exam, write down the rounded values for each variable, and show each step of your working.

2. A racetrack measures 800 m to the nearest 10 metres. A runner can run around the track in 178 s to the nearest second. Calculate the runner's fastest speed to a suitable degree of accuracy, and explain your answer.

   Underline the key information and the instruction.

   upper bound of distance: 805 m
   lower bound of distance: 795 m
   upper bound of time: 178.5 s
   lower bound of time: 177.5 s

   Calculate the upper and lower bounds of the values.

   $\text{speed} = \frac{\text{distance}}{\text{time}}$

   upper bound of speed $= \frac{805}{177.5} = 4.53...$ m/s

   lower bound of speed $= \frac{795}{178.5} = 4.45...$ m/s

   speed $= 4.5$ m/s to 2 s.f., because the upper and lower bounds of the speed are both 4.5 m/s to 2 s.f.

   To calculate the fastest speed, you want the largest possible solution from the division, so use $\frac{\text{upper bound of distance}}{\text{lower bound of time}}$.

Estimation

When **estimating**, round numbers to 1 significant figure before performing the calculation.

Worked example

Estimate the value of $\frac{8.762 \times 13.7}{4.82}$

$\frac{8.762 \times 13.7}{4.82}$

$\approx \frac{9 \times 10}{5}$ — Rewrite the calculation with all numbers rounded to 1 s.f.

$= 18$

$\approx$ means 'approximately equal to'.

Retrieval 2

2 Rounding, truncating, error intervals and estimating

Learn the answers to the questions below, then cover the answers column with a piece of paper and write as many as you can. Check and repeat.

	Questions	Answers
1	What is rounding?	Writing a number to a given place value.
2	If the value you are considering for rounding is 5, do you round up or down?	Round up.
3	When you round to 1 decimal place, which number tells you whether to round up or round down?	The number in the second decimal place.
4	What is the first significant figure of a number?	The first non-zero digit, counting from the left.
5	In the number 56 743, which is the 4th significant figure?	4 (round to the nearest 10).
6	In the number 0.000 3654, which is the 2nd significant figure?	6 (round to the nearest 0.000 01).
7	How do you find the error interval if a number has been rounded to a given unit?	Subtract half a unit for the minimum and add half a unit for the maximum.
8	How do you find the error interval if a number has been truncated to a given unit?	The minimum is the given measurement, add one unit for the maximum.
9	When multiplying error intervals, what calculation should you use to find the maximum possible value?	upper bound × upper bound = maximum value
10	What is an error interval?	An inequality that shows the range of possible values of unrounded numbers.

Previous questions

Now go back and use these questions to check your knowledge of previous topics.

	Questions	Answers
1	What does commutative mean?	Can be done in any order.
2	What is another word for the result of raising something to the power of −1?	Reciprocal.
3	What is the quotient of two numbers?	The result you get when you divide one of the numbers by the other.
4	What is the value of $(xy)^0$?	1

2 Retrieval 13

Exam-style questions

2.1 A number, P, is truncated to 1 digit. The result is 1

A number, Q, is truncated to 1 decimal place. The result is 0.6

Write down the error interval for $P + Q$. [3 marks]

2.2 $\sqrt{6} = 2.449$ correct to 4 significant figures.

Find the value of $\sqrt{0.06}$ correct to 3 significant figures. [3 marks]

2.3 $A = \dfrac{B}{C^2}$

$B = 6.2$ correct to 2 significant figures

$C = 20.3$ correct to 3 significant figures

(a) Calculate the lower bound of A to 5 significant figures. [3 marks]

The upper bound for A is 0.015242 to 5 significant figures.

(b) By considering bounds, write down the value of A to a suitable degree of accuracy.

You must give a reason for your answer. [2 marks]

2.4 A steel block is made in the shape of a cuboid.

The length of the block is 30.0 cm to the nearest mm.

The width of the block is 25.4 cm to the nearest mm.

The height of the block is 9.2 cm to the nearest mm.

The mass of the block, to the nearest 250 g, is 56.5 kg.

By considering bounds, show that the density of the steel is 0.008 kg/cm³, correct to one significant figure.

Show all working and give reasons for your final answer. [5 marks]

2.5 Round 20 193 to

(a) 4 significant figures [1 mark]

(b) 3 significant figures [1 mark]

(c) 2 significant figures [1 mark]

(d) 1 significant figure. [1 mark]

2.6 (a) Use your calculator to work out $\frac{1}{3}(0.02 \times 11.9)^2$.

Write all the figures on your calculator display. [1 mark]

(b) Write your answer to part (a)

(i) truncated to 2 decimal places [1 mark]

(ii) rounded to 2 significant figures. [1 mark]

2.7 Akira rounds 0.065 29 to 2 significant figures and gives the answer 0.07 Akira is wrong. Explain why. [1 mark]

2.8 Work out an estimate for the value of $\frac{2.67 \times 1.36}{0.11 + 0.42}$ [2 marks]

2.9 A biologist visits a lake at the start of January and works out that the number of fish in the lake is approximately 1000. She thinks that the population is growing at a rate of 17 fish per day. Estimate how many fish there will be in the lake five months later. [3 marks]

2.10 In one week, an Italian restaurant sells 96 portions of lasagne. The restaurant sells a portion of lasagne for £8.95 and each portion costs £3.20 to make. Estimate the profit the restaurant makes from lasagne in the week. [3 marks]

2.11 Work out an estimate for the value of each square root. Give your answers to 1 decimal place.

(a) $\sqrt{47}$ [1 mark]

(b) $\sqrt{200}$ [1 mark]

Questions referring to previous content

2.12 Write $\frac{\sqrt{5^3}}{5} + 2\sqrt{125}$ in index form, simplified as much as possible. [4 marks]

2.13 Put one or more sets of brackets in the following statement to make it true:

$20 \div 12 - 7 + 8 \times 4 + 1 + 3 = 47$ [2 marks]

Knowledge

3 Factors, multiples, primes, standard form, and surds

Factors, multiples, and prime numbers

A **factor** is an **integer** that divides exactly into a number.

3 and 4 are factors of 12.

A multiple is the result obtained when one integer is multiplied by another.

12 is a **multiple** of both 3 and 4.

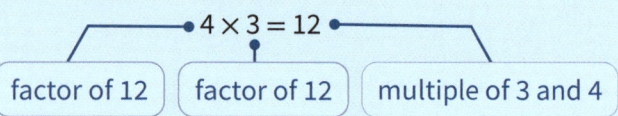

For example, multiples of 5 are:

$1 \times 5 = 5$

$2 \times 5 = 10$

$3 \times 5 = 15$... and so on.

For large numbers we can use **prime factor decomposition** to help us find the HCF and LCM. A **prime number** is a number with exactly two factors. A **prime factor** is a factor that is also a prime number.

> The **highest common factor** (**HCF**) of two numbers is the *biggest* number that is a factor of both of the numbers.
>
> The **lowest common multiple** (**LCM**) of two numbers is the *smallest* number that is a multiple of both of the numbers.

WATCH OUT

1 is not a prime number as it only has one factor (which is 1).

Worked example

Write 189 as a product of its prime factors.

$189 = 3 \times 3 \times 3 \times 7$

$189 = 3^3 \times 7$

Use a prime factor tree to write a number as a product of its prime factors.

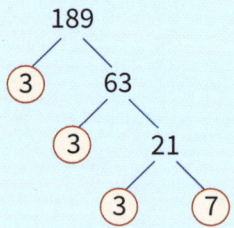

Note, any correct factor pairs will give the same answer.

Factors, multiples, and prime numbers

Worked example

Find the HCF and LCM of 150 and 225.

$150 = 2 \times 3 \times 5 \times 5$

$225 = 3 \times 3 \times 5 \times 5$

Write 150 and 225 as products of their prime factors.

REVISION TIP

Use a factor tree to find the prime factors.

Draw a Venn diagram of the factors.

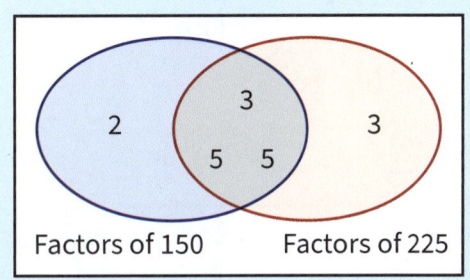

Factors of 150   Factors of 225

$HCF = 3 \times 5 \times 5 = 75$

HCF is the product of numbers in the intersection.

$LCM = 2 \times 3 \times 5 \times 5 \times 3 = 450$

LCM is the product of all the numbers.

Worked example

Two numbers have prime factor decomposition $2^2 \times 3 \times 5^2$ and $2^3 \times 5 \times 11^2$

(a) Find the highest common factor.
Leave your answer in index form.

$HCF = 2 \times 2 \times 5$
$ = 2^2 \times 5$

Find the prime factors and show them in a Venn diagram.

(b) Find the lowest common multiple.
Leave your answer in index form.

$LCM = 5 \times 3 \times 2 \times 2 \times 5 \times 2 \times 11 \times 11$
$ = 2^3 \times 3 \times 5^2 \times 11^2$

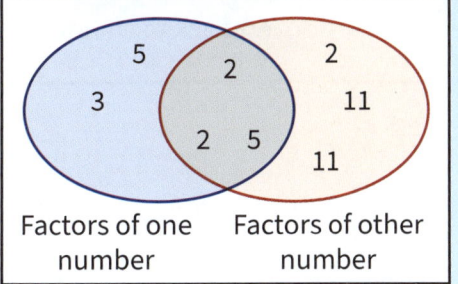

Factors of one   Factors of other
number   number

3 Knowledge

Knowledge

3 Factors, multiples, primes, standard form, and surds

Standard form: The basics

Standard form is when a power of 10 is used to rewrite a very large or very small number.

Standard form is written in this way:

[between 1 and 10 i.e. $1 \leq a < 10$] ⟶ $a \times 10^n$ ⟵ [positive or negative integer (or 0)]

LINK
- To remind yourself about multiplying and dividing by powers of 10, look back at Chapter 1.
- To remind yourself about indices, look back at Chapter 2.

These numbers are not written in standard form.

[This number needs to lie in the interval $1 \leq a < 10$] ⟶ 16×10^7

$2.4 \times 10^{0.5}$ ⟵ [This number needs to be an integer]

Adding or subtracting numbers in standard form

To add or subtract numbers in standard form, convert them to ordinary numbers first.

Worked example

Calculate the value of $8.9 \times 10^{-4} - 6 \times 10^{-6}$

Give your answer in standard form.

$8.9 \times 10^{-4} = 0.00089$
$6 \times 10^{-6} = 0.000006$ ⟵ [Convert to ordinary numbers.]

0.000890
-0.000006 ⟵ [Subtract]
$\overline{0.000884}$

$= 8.84 \times 10^{-4}$ ⟵ [Convert back to standard form.]

LINK
To remind yourself about addition and subtraction, look back at Chapter 1.

Multiplying or dividing numbers in standard form

To multiply or divide numbers in standard form, you need to use these two laws of indices:

1. $a^n \times a^m = a^{n+m}$ If you multiply numbers with the same base, you add the powers.
2. $a^n \div a^m = a^{n-m}$ If you divide numbers with the same base, you subtract the powers.

LINK

To remind yourself about laws of indices, look back at Chapter 2.

Worked example

Given that $a = 6 \times 10^5$ and $b = 3 \times 10^{-2}$, work out the value of:

1. $a \times b = (6 \times 10^5) \times (3 \times 10^{-2})$
   $= (6 \times 3) \times (10^5 \times 10^{-2})$ — Group ordinary numbers on one side.
   $= 18 \times 10^{5+(-2)}$ — Use $a^n \times a^m = a^{n+m}$ to simplify the indices.
   $= 18 \times 10^3$
   $= 1.8 \times 10^4$ — Write your answer in standard form.

2. $a \div b = (6 \times 10^5) \div (3 \times 10^{-2})$
   $= \dfrac{6 \times 10^5}{3 \times 10^{-2}}$ — Write as a fraction.
   $= \dfrac{6}{3} \times \dfrac{10^5}{10^{-2}}$ — Group ordinary numbers and powers of 10.
   $= 2 \times 10^{5-(-2)}$ — Use $a^n \div a^m = a^{n-m}$
   $= 2 \times 10^7$ — Write your answer in standard form.

In the previous Worked example, part 1, you saw that 18×10^3 is **not** in standard form. To make the first number between 1 and 10, you must divide 18 by 10 and multiply 10^3 by 10.

Worked example

Write these numbers in standard form.

1. 23×10^4

   23×10^4
   $\div 10 \quad \times 10$
   2.3×10^5

   — To make the first number between 1 and 10, divide 23 by 10.
   — Multiply by 10 to keep the same value.

2. 0.023×10^4

   0.023×10^4
   $\times 100 \quad \div 100$
   2.3×10^2

   — To change 0.023 to 2.3, multiply by 100.
   — Divide by 100 to keep the same value.

3 Knowledge

Knowledge

3 Factors, multiples, primes, standard form, and surds

Surds

A **surd** is an irrational **square root**. Surds can often be simplified using these two laws:

$\sqrt{ab} = \sqrt{a}\sqrt{b}$ and $\sqrt{\dfrac{a}{b}} = \dfrac{\sqrt{a}}{\sqrt{b}}$

Worked example

Simplify fully $\sqrt{45} - \sqrt{32} + \sqrt{5}$

$\sqrt{45} = \sqrt{9}\sqrt{5}$
$\phantom{\sqrt{45}} = 3\sqrt{5}$
$\sqrt{32} = \sqrt{16}\sqrt{2}$
$\phantom{\sqrt{32}} = 4\sqrt{2}$
$\sqrt{45} - \sqrt{32} + \sqrt{5}$
$= 3\sqrt{5} - 4\sqrt{2} + \sqrt{5}$
$= 4\sqrt{5} - 4\sqrt{2}$

- Simplify each surd. Look for a factor that is a square number. ($\sqrt{5}$ can't be simplified)
- Rewrite the calculation using the simplified surds.
- Collect like terms.
- This can't be simplified any further.

Rewriting a fraction to remove any surds from the denominator is called rationalising the denominator.

Worked example

1. Rationalise the denominator and rewrite $\dfrac{1}{2 + \sqrt{3}}$

$\dfrac{1(2 - \sqrt{3})}{(2 + \sqrt{3})(2 - \sqrt{3})}$

$= \dfrac{2 - \sqrt{3}}{4 - 2\sqrt{3} + 2\sqrt{3} - 3}$

$= \dfrac{2 - \sqrt{3}}{1}$

$= 2 - \sqrt{3}$

- There are **two terms** in the denominator, so multiply the numerator and denominator by the denominator with the *opposite sign* between the terms.
- Expand the brackets.
- These cancel each other out.
- Multiply the numerator and denominator by the denominator with the opposite sign between the terms.

2. Express $\dfrac{1 + \sqrt{8}}{1 - \sqrt{2}}$ in the form $a + b\sqrt{2}$, where a and b are integers.

$\dfrac{1 + \sqrt{8}}{1 - \sqrt{2}} = \dfrac{(1 + \sqrt{8})(1 + \sqrt{2})}{(1 - \sqrt{2})(1 + \sqrt{2})}$

$= \dfrac{1 + \sqrt{2} + \sqrt{8} + \sqrt{16}}{1 - 2}$

$= \dfrac{1 + \sqrt{2} + 2\sqrt{2} + 4}{-1}$

$= -5 - 3\sqrt{2}$

- Expand the brackets.
- Simplify any surds.
- Collect like terms and simplify.
- An expression over -1 becomes the negative of the expression.

Key terms — Make sure you can write a definition for these key terms.

factor   highest common factor   lowest common multiplier
multiple   prime factor   prime factor decomposition
prime number   root   standard form   surd

Retrieval 3

3 Factors, multiples, primes, standard form, and surds

Learn the answers to the questions below, then cover the answers column with a piece of paper and write as many as you can. Check and repeat.

Questions | Answers

1. What is a factor? — An integer that divides exactly into another integer leaving no remainder.
2. What is a multiple? — A multiple of a is $n \times a$, where n is any other integer.
3. What do HCF and LCM stand for? — Highest common factor and Lowest common multiple.
4. What is a prime number? — A number with only two factors, one and itself.
5. Why is 2 the only even prime number? — All other even numbers are divisible by 2 and therefore have more than two factors.
6. When a number is written in standard form, what form does it take? — $a \times 10^n$, where $1 \leq a < 10$ and n is an integer
7. When a number is written in standard form ($a \times 10^n$) what values can the a take? — $1 \leq a < 10$
8. When a number is written in standard form ($a \times 10^n$) what values can the index number, n, take? — Any integer.
9. What is standard form used for? — To write very large or very small numbers.
10. How do you add/subtract numbers that are in standard form? — Convert them to ordinary numbers, add/subtract them and then convert back into standard form.
11. What is a surd? — An irrational square root.
12. How do you simplify a surd? — Look for a factor that is a square number.
13. How do you rationalise the denominator where there is a single surd in the denominator? — Multiply the numerator and denominator by the surd in the denominator.

Previous questions

Now go back and use these questions to check your knowledge of previous topics.

Questions | Answers

1. What is an integer? — A whole number.
2. What does the letter B in BIDMAS stand for? — Brackets.
3. What is the first significant figure of a number? — The first non-zero digit, counting from the left.
4. What does $(a^n)^m$ simplify to? — a^{nm}.

Practice

Exam-style questions

3.1 Express each of these in the form $a\sqrt{2}$, where a is an integer.

(a) $\sqrt{200} + \sqrt{72} - \sqrt{98}$ [2 marks]

(b) $\dfrac{14}{\sqrt{2}}$ [2 marks]

3.2 $\sqrt{3}(\sqrt{80} - \sqrt{20})$ can be written in the form $a\sqrt{15}$, where a is an integer. Find the value of a. [3 marks]

3.3 Express each of these in the form $a + b\sqrt{3}$, where a and b are integers.

(a) $\dfrac{8}{2 - \sqrt{3}}$ [3 marks]

(b) $\dfrac{\sqrt{3} - 1}{\sqrt{3} + 1}$ [3 marks]

3.4 Show that $(\sqrt{11} - \sqrt{8})(\sqrt{11} + \sqrt{8}) = 3$ [3 marks]

3.5 Show that $\dfrac{(\sqrt{2} + \sqrt{50})^2}{3(\sqrt{2} - 1)}$ can be written in the form $a(b + \sqrt{2})$, where a and b are integers. [3 marks]

3.6 (a) Write 750 as a product of its prime factors. Give your answer in index notation. [2 marks]

(b) By looking at its prime factors, explain why 750 is not divisible by 4. [1 mark]

3.7 The prime factor decomposition of a number, x, is $2 \times 3^2 \times 7 \times 13$

(a) Is x even or odd? Explain your reasoning. [1 mark]

(b) Write down the prime factor decomposition for $2x$. [1 mark]

3.8 $P = 2^3 \times 5 \times 11 \qquad Q = 2 \times 3^2 \times 5$

(a) Write down the highest common factor (HCF) of P and Q. [1 mark]

(b) Write down the lowest common multiple (LCM) of P and Q. [1 mark]

3.9 $p = 2 \times 3 \times 7 \times x$ $\qquad$ $q = 2^2 \times 5^2$ $\qquad$ $x > 0$

The highest common factor of p and q is 4.

(a) Work out the lowest possible value of x. [1 mark]

(b) Work out the lowest possible value of p. [1 mark]

EXAM TIP
Think about how the HCF links to the prime factor decomposition.

3.10 The distance from the Sun to Earth is approximately 150 000 000 km. Write this number in standard form. [1 mark]

3.11 Write these numbers in order of size, starting with the greatest.

2.1×10^4 $\qquad$ 2.3×10^5 $\qquad$ 0.21×10^4 $\qquad$ 2200 [3 marks]

EXAM TIP
Write all the numbers in the same form.

3.12 Work out the value of each expression, giving your answers in standard form.

(a) $(5 \times 10^4) + (6 \times 10^5)$ [2 marks]

(b) $(9 \times 10^{-3}) - (3 \times 10^{-4})$ [2 marks]

(c) $(2.1 \times 10^8) \times (3 \times 10^{-5})$ [2 marks]

(d) $(8.2 \times 10^3) \div (4.1 \times 10^7)$ [2 marks]

3.13 A jet can fly at 4×10^3 km/h. How long would it take to travel a distance of 3000 km? Give your answer in minutes. [3 marks]

3.14 A region on a map forms the shape of a rectangle with width 1.2×10^2 km and length 7×10^3 km. Work out the area of this region. Give your answer in standard form. [3 marks]

3.15 (a) Work out $(3.02 \times 10^{18}) \times (6.3 \times 10^{-8})$

Give your answer as an ordinary number. [2 marks]

(b) Work out $\dfrac{5.3 \times 10^8 - 8.9 \times 10^{-2}}{7.2 \times 10^3}$

Give your answer in standard form. [2 marks]

Questions referring to previous content

3.16 Which is greater, and by how much: [3 marks]

the sum of $21.263 + 4.801$, rounded to 3 significant figures,

or the difference of $88.155 - 61.379$, truncated to a whole number?

3.17 Estimate the value of the following expression, by first rounding each number to 1 significant figure:

$\dfrac{4.2^2 \times 5.075}{9.88} + \dfrac{11.7}{2.66}$ [2 marks]

Knowledge

4 Fractions, decimals, percentages

Fractions

The **numerator** is the top number in a **fraction**.

The **denominator** is the bottom number in a fraction.

$\dfrac{3}{5}$ — numerator / denominator

An **improper fraction** has a numerator larger than or equal to the denominator. For example, $\dfrac{15}{7}$

A **mixed number** has a whole part and a fraction part. For example, $2\dfrac{1}{3}$

Worked example

1. Write $4\dfrac{3}{7}$ as an improper fraction.

   $4 = \dfrac{4 \times 7}{7}$ — Convert the whole part into an improper fraction by multiplying by the denominator.

   $= \dfrac{28}{7}$

   $= \dfrac{28}{7} + \dfrac{3}{7}$ — Add the answer for the whole part to the fraction part.

   $= \dfrac{31}{7}$

2. Write $\dfrac{17}{6}$ as a mixed number.

   $\dfrac{17}{6} = 2$ remainder 5 — Divide the numerator by the denominator.

   $= 2\dfrac{5}{6}$ — Write this as a whole number and a fraction.

To compare fractions with different denominators, convert the fractions to **equivalent fractions** that have a common denominator or common numerator.

Worked example

Put these fractions in ascending order: $\dfrac{2}{3}, \dfrac{5}{12}, \dfrac{5}{8}$

$3 = 1 \times 3$

$12 = 3 \times 2 \times 2$ — Find the LCM of 3, 12, and 8.

$8 = 2 \times 2 \times 2$

LCM $= 2 \times 2 \times 2 \times 3$

$= 24$

> **LINK**
> To remind yourself about the HCF and LCM, look back at Chapter 3.

$\dfrac{2}{3} = \dfrac{16}{24}$

$\dfrac{5}{12} = \dfrac{10}{24}$ — Convert the fractions to 24ths by identifying the number the denominator needs to be multiplied by to get to 24, and multiplying the numerator by the same number.

$\dfrac{5}{8} = \dfrac{15}{24}$

$\dfrac{10}{24}, \dfrac{15}{24}, \dfrac{16}{24}$ — Put the fractions in order by comparing the numerators (smallest first).

$\dfrac{5}{12}, \dfrac{5}{8}, \dfrac{2}{3}$ — Write the fractions in their original form in the order corresponding to the previous step.

Calculations with fractions

Multiplying fractions

Multiply the numerators and multiply the denominators.

- $\dfrac{4}{5} \times \dfrac{2}{3} = \dfrac{2 \times 4}{5 \times 3} = \dfrac{8}{15}$

Worked example

Calculate these and give your answers in their simplest form.

(a) $\dfrac{3}{4}$ of $\dfrac{5}{6}$

$\dfrac{3}{4}$ of $\dfrac{5}{6} = \dfrac{3}{4} \times \dfrac{5}{6}$

$= \dfrac{15}{24}$ — Simplify the fraction.

$= \dfrac{5}{8}$

(b) $2\dfrac{7}{9} \times \dfrac{3}{10}$

$2\dfrac{7}{9} \times \dfrac{3}{10} = \dfrac{25}{9} \times \dfrac{3}{10}$ — Convert $2\dfrac{7}{9}$ to an improper fraction.

$= \dfrac{75}{90}$

$= \dfrac{5}{6}$

> **REVISION TIP**
> After all fraction calculations, use equivalent fractions to simplify your answer if possible.

Dividing fractions

To divide by a fraction, multiply by its reciprocal.

$\dfrac{4}{5} \div \dfrac{2}{3} = \dfrac{4}{5} \times \dfrac{3}{2} = \dfrac{4 \times 3}{5 \times 2} = \dfrac{12}{10}$

Worked example

Calculate these and give your answers in their simplest form.

(a) $\dfrac{3}{8} \div \dfrac{7}{10}$

$\dfrac{3}{8} \div \dfrac{7}{10} = \dfrac{3}{8} \times \dfrac{10}{7}$ — Swap to a multiplication by changing the sign and swapping the numerator and denominator of the second fraction.

$= \dfrac{30}{56}$

$= \dfrac{15}{28}$ — Multiply the numerators and the denominators, and simplify.

(b) $5 \div \dfrac{1}{3}$

$5 \div \dfrac{1}{3} = 5 \times 3$

$= 15$

> **REVISION TIP**
> The **reciprocal** of a number is the number you would have to multiply it by to get the answer 1. For example, the reciprocal of 2 is $\dfrac{1}{2}$.

Adding and subtracting fractions

Change any mixed numbers into improper fractions. Write both fractions over a common denominator. Add or subtract the numerators.

$\dfrac{4}{5} + \dfrac{2}{3} = \dfrac{12}{15} + \dfrac{10}{15} = \dfrac{22}{15}$

Worked example

1. Two pipes of lengths $2\dfrac{3}{4}$ m and $1\dfrac{2}{3}$ m are joined end to end. How long is the new length of pipe?

$2\dfrac{3}{4} m + 1\dfrac{2}{3} m = \dfrac{11}{4} m + \dfrac{5}{3} m$ — Rewrite mixed numbers as improper fractions.

$= \dfrac{33}{12} m + \dfrac{20}{12} m$ — Find the LCM of the denominators (12) and write the fractions with this as the new denominator.

$= \dfrac{53}{12} m$

$= 4\dfrac{5}{12} m$ — Add fractions and convert answer to a mixed number.

> **WATCH OUT**
> It's usually easiest to write mixed numbers as improper fractions before you add or subtract them.

4 Knowledge

Knowledge

4 Fractions, decimals, percentages

Converting between fractions, decimals, and percentages

Percent means "out of 100".

Percentages can be written as fractions with 100 on the denominator.

Percentages can be converted into decimals or fractions, and vice versa.

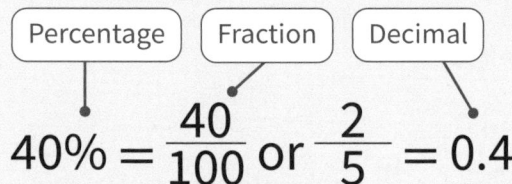

$$40\% = \frac{40}{100} \text{ or } \frac{2}{5} = 0.4$$

Worked example

Niamh has some money. She spends $\frac{2}{5}$ on shoes and 35% on clothes.

What fraction of the original amount does she have left?

$35\% = \frac{35}{100}$ — Convert 35% to a fraction.

$\frac{2}{5} = \frac{40}{100}$ — Write $\frac{2}{5}$ as an equivalent fraction with denominator 100.

$\frac{35}{100} + \frac{40}{100} = \frac{75}{100}$ has been spent. — Add the fractions to find the total amount spent.

$1 - \frac{75}{100} = \frac{100}{100} - \frac{75}{100}$ — Subtract the fraction spent from 1 to find the fraction that is left.

$= \frac{25}{100}$

$= \frac{1}{4}$ — Simplify the fraction.

Worked example

There are 16 balls in a bag and 5 of them are red. What percentage of the balls are red?

$\frac{5}{16}$ of the balls in the bag are red. — Write the fraction of red balls in the bag.

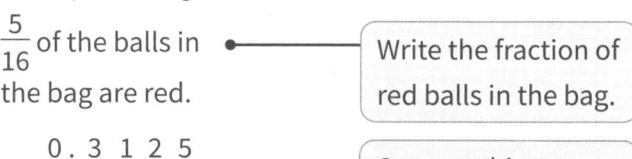

— Convert this to a decimal by dividing.

$0.3125 \times 100 = 31.25$ — Multiply by 100 to convert to a percentage.

31.25% of balls are red.

A recurring decimal is a decimal number which has a repeating number or pattern of numbers after the decimal point. A dot over the recurring numbers show they are recurring.

$5.333333... = 5.\dot{3}$ $\qquad\qquad$ $3.717171... = 3.\dot{7}\dot{1}$

Recurring decimals can be converted into fractions.

4

Converting between fractions, decimals, and percentages

Worked example

Write $0.\dot{7}\dot{2}$ as a fraction.

Let $x = 0.7272\ldots$
$100x = 72.7272\ldots$ — Two digits recur, so multiply by 100.
$100x - x = 72.7272\ldots - 0.7272\ldots$
$99x = 72$
$x = \dfrac{72}{99}$ — Clear the decimal places by subtracting x from $100x$.
$x = \dfrac{8}{11}$

Divide both sides by 99 to find x.

REVISION TIP

If only the first digit was recurring, you'd multiply by 10 and subtract to give $9x$.

Worked example

Prove algebraically that $0.5\dot{8} = \dfrac{53}{90}$

Let $x = 0.5888\ldots$
$10x = 5.888\ldots$
$100x = 58.888\ldots$
$100x - 10x$
$= 58.888\ldots - 5.888\ldots$
$90x = 53$
$x = \dfrac{53}{90}$

Start by multiplying $0.5888\ldots$ by 10 so only the recurring digits appear after the decimal point.

Multiply by 100, and subtract $90x$ to give $10x$, then solve.

WATCH OUT

This example is slightly different because the first digit after the decimal point (5) doesn't recur.

Percentages

Percentages of amounts can be calculated using **multipliers**.

original amount × multiplier = new amount

Worked example

Calculate 63% of 576.

$63\% = \dfrac{63}{100}$ — Find the decimal equivalent of 63%
$= 0.63$
$= 576 \times 0.63$ — Multiply the amount by the multiplier (you can use a calculator for this).
$= 362.88$

Amounts can be increased or decreased by a given percentage.

Worked example

1. Increase 75 by 8%

   $100\% + 8\% = 108\%$ — Add the percentage increase to 100%
   $108\% = \dfrac{108}{100}$
   $= 1.08$ — Find the multiplier.
   $75 \times 1.08 = 81$ — Multiply the original amount by the multiplier.

2. Decrease 75 by 8%

   $100\% - 8\% = 92\%$ — Subtract the percentage from 100%
   $92\% = \dfrac{92}{100}$
   $= 0.92$ — Find the multiplier.
   $75 \times 0.92 = 69$ — Multiply the original amount by the multiplier.

4 Knowledge

Knowledge

4 Fractions, decimals, percentages

Percentages

Worked example

The price of an armchair is £52.34 including VAT.
Calculate the original price of the armchair before VAT.

$52.34 = 120\%$ of x — Let £x be the original price. Find an expression for x.

$52.34 = 1.2x$

$x = \dfrac{52.34}{1.2}$

$x = 43.61666\ldots$ — Rewrite the expression using a multiplier ($120\% = 1.2$) and solve for x.

original price = £43.62

REVISION TIP
VAT is 20% of the original price. It is added on to the cost of the item.

You might be given an amount before and after a change and asked to calculate the percentage increase or decrease.

$$\text{percentage increase} = \left(\dfrac{\text{actual increase}}{\text{original amount}}\right) \times 100\% \qquad \text{percentage decrease} = \left(\dfrac{\text{actual decrease}}{\text{original amount}}\right) \times 100\%$$

Worked example

The height of a plant increases from 64 cm to 87.04 cm.
Calculate the percentage increase in the height of the plant.

Method 1: Using the formula

$87.04 - 64 = 23.04$ cm — Subtract the starting amount from the end amount.

percentage increase = $\dfrac{23.04}{64} \times 100\%$

$= 0.36 \times 100\%$

$= 36\%$

Divide the difference by the starting amount and multiply by 100 to find the percentage increase.

Method 2: Using multipliers

$64x = 87.04$ — Write an equation that includes the start and end amounts with x as the multiplier.

$x = \dfrac{87.04}{64}$

$x = 1.36$, so the percentage increase in height is 36%

Rearrange to find the value of x.

REVISION TIP
A 1 before the decimal point = an increase. (original amount plus a bit extra)

A 0 before the decimal point = a decrease (less than the original amount)

Percentages

To increase or decrease an amount, A, by the same percentage n times, use
$A \times \text{multiplier}^n$

This is used to calculate **compound interest**. This is where the interest is left in the account, so interest is earned on the original amount plus the interest.

The other type of interest is **simple interest**. This is where you only earn interest on the initial investment, or the interest is taken out at the end of each year.

> **Worked example**
>
> Natalie keeps £3500 in a savings account for five years at an interest rate of 4% per year. She needs to decide whether to take the interest out of the account each year or leave it in. Calculate how much extra interest she will earn if she leaves it in for the full five years.
>
> 4% of £3500 = 0.04 × £3500
>   = £140
>
> £140 × 5 = £700
>
> total after 5 years = 3500×1.04^5
>   = £4258.29
>
> £4258.29 − £3500 = £758.29
>
> £758.29 − £700 = £58.29
>
> - If the interest is taken out at the end of each year, it is **simple interest**. Calculate the interest earned in one year.
> - Multiply by 5 to calculate the amount of simple interest earned over five years.
> - If the interest is left in, it is **compound interest**. Convert the interest percentage to a multiplier (4%, so 1.04) Use the number of years as the power. Substitute into the compound interest expression to find the total in the account after five years.
> - To find the interest earned, subtract the initial investment.
> - Subtract the simple interest amount from the compound interest amount to find the difference.

Key terms — Make sure you can write a definition for these key terms.

compound interest   denominator   equivalent fraction   fraction   improper fraction   multiplier   numerator   percent   percentage   reciprocal   simple interest

Retrieval

4 Fractions, decimals, percentages

Learn the answers to the questions below, then cover the answers column with a piece of paper and write as many as you can. Check and repeat.

	Questions	Answers
1	What is an improper fraction?	A fraction where the numerator is larger than or equal to the denominator.
2	What is a mixed number?	A number that has a whole number and a fraction.
3	What is the reciprocal of a number?	The reciprocal of a number is 1 over the number. It is the value that you multiply a number by to get 1.
4	How do you multiply two fractions?	Multiply the numerators together and multiply the denominators together.
5	How do you divide by a fraction?	Multiply by its reciprocal.
6	How do you add or subtract fractions?	Write all the fractions with common denominators and add or subtract the numerators.
7	How do you add or subtract mixed numbers?	Change into improper fractions, then write all the fractions with common denominators, then add the numerators.
8	What form should you leave a fractional answer in?	As a fully simplified proper fraction or mixed number.
9	How can you find the fraction of an amount?	Multiply the fraction by the amount.
10	How do you convert a percentage into a decimal?	Divide the percentage by 100.
11	How do you change a percentage into a fraction?	Write the percentage as a fraction with 100 as the denominator.
12	What is simple interest?	A fixed amount of interest calculated as a percentage of the original amount.
13	What is compound interest?	An amount of interest that is calculated as a percentage of the current amount.

Previous questions

Now go back and use these questions to check your knowledge of previous topics.

	Questions	Answers
1	What is rounding?	Writing a number to a given place value.
2	Describe how you would simplify $a^5 \div a^4$	Subtract the powers.
3	What does BIDMAS tell you?	The order in which to do a calculation.

Exam-style questions

4.1 Write the following numbers in descending order:

16% $\frac{1}{6}$ 0.165 $\frac{17}{100}$ **[2 marks]**

4.2 A caravan costs £26 759 plus VAT at 20%.

The purchaser needs to pay a deposit up front.

The remainder is then paid in 18 equal payments of £1450.60

Work out how much the purchaser needs to pay for the deposit.   **[4 marks]**

4.3 Maxwell is reading a book on his e-reader. When he picks it up one day, it tells him he is $\frac{1}{3}$ of the way through the book. He reads some and when he puts it down he is $\frac{3}{4}$ of the way through. What fraction of the book did he read?

[2 marks]

4.4 Work out the perimeter of the shape shown.

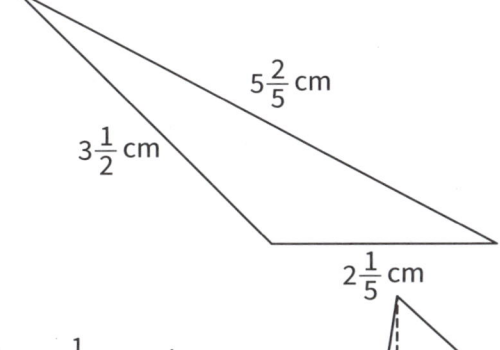

[3 marks]

4.5 A triangle has base $1\frac{1}{5}$ cm and height $\frac{6}{5}$ cm.

A rectangle has width $\frac{2}{5}$ cm and length x cm.

The rectangle and the triangle have the same area.

Find x.

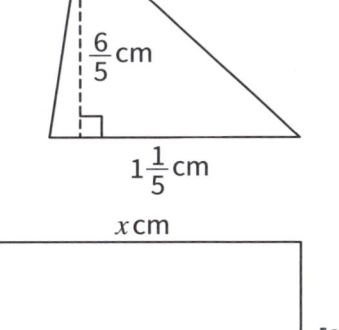

[3 marks]

Exam-style questions

4.6 Vasiliki has a piece of material $3\frac{3}{4}$ m long.

She is cutting it into smaller pieces of length $\frac{5}{6}$ m.

How many smaller pieces can she get, and what fraction of a metre will be left over?

Number of small pieces = _____

Fraction left = _____ [2 marks]

4.7 Autumn ran some park races last year. 15% of her races were 5 km runs, $\frac{7}{10}$ of her races were 10 km runs and the rest were half marathons. She ran 20 races in total.

How many were half marathons? [3 marks]

4.8 Express these fractions as decimals.

(a) $\frac{1}{18}$ (b) $\frac{20}{33}$ [2 marks]

4.9 Prove algebraically that $0.\dot{5} = \frac{5}{9}$ [2 marks]

EXAM TIP

Let $x = 0.\dot{5}$ and find $10x$

4.10 Write $0.6\dot{4}$ as a fraction in its simplest form. [2 marks]

4.11 Prove algebraically that $0.0\dot{5}\dot{6} = \frac{28}{495}$ [3 marks]

4.12 A baby lizard is 6 cm long.

When it is adult, the lizard is 22 cm long.

Work out the percentage increase in the length of the lizard.

Give your answer to 3 significant figures. [3 marks]

4.13 Darren runs a marathon in 3 hours and 15 minutes.

After some training, he runs it in 3 hours.

Calculate the percentage decrease in the time it takes Darren to run a marathon. [2 marks]

EXAM TIP

Convert the times into minutes first.

4.14 Farida pays £150 for a box of 10 000 screws.

She packs them into smaller boxes containing 1250 screws per box.

She sells all of these boxes for £24 per box.

Work out Farida's percentage profit. [4 marks]

4.15 The price of a car was increased by 50% and then reduced by 25%.

Work out the overall percentage increase in the original price. **[3 marks]**

 4.16 The numerator of a fraction is increased by 48%.

Its denominator is decreased by $87\frac{1}{2}\%$.

The resulting fraction is $\frac{37}{42}$

Give the original fraction in its simplest form. **[3 marks]**

 4.17 Two banks have an interest rate of 6% per annum. Bank A gives compound interest and Bank B gives simple interest.

(a) In January 2023, Keysha invests £2450 in Bank A and Paul invests £2450 in Bank B.

How much more money will Keysha have than Paul in January 2030? **[4 marks]**

(b) Phoebe invests money in Bank A. She needs at least £5000 in her account by January 2033. Work out how much, in whole pounds, Phoebe needs to invest in January 2023 to reach this figure. **[3 marks]**

 4.18 Avani invests £6000 in an account for one year. At the end of the year, interest is added to her account.

Avani pays tax on the interest at a rate of 15%. She pays £19.80 tax.
Work out the percentage interest rate for the account. **[3 marks]**

 4.19 The value of an oil painting is increasing at a rate of x% per year.

Initially, the painting is worth £4000. After five years, it is worth £4300
Work out the value of x to 1 decimal place. **[3 marks]**

 4.20 Ehmet bought a flat. Its value fell by 3% in the first year. During the next two years, the value went up by 7% per year. At the end of the third year, his flat was worth £285 000

(a) Calculate the percentage change in the value of the flat to 3 significant figures. **[3 marks]**

(b) Work out how much Ehmet paid for his flat. **[2 marks]**

Questions referring to previous content

 4.21 By drawing a Venn diagram, or by any other method, find the HCF and LCM of 300 and 840. **[4 marks]**

 4.22 Show that $(\sqrt{3} + 2\sqrt{27})^2 = 147$ **[4 marks]**

Knowledge

5 Algebra basics

Algebraic expressions

Variable A letter in an expression or equation, such as x or y

Term A number or letter by itself, or multiple numbers and letters multiplied together, for example, $3x, -\frac{y}{2}, 2ab^2, -3$

Expression A collection of letters and numbers with no equals or inequality sign, such as $2x + 3y - 4$

Equation Two expressions that are connected by an equals sign, such as $2x + 4 = 3x + 1$

Simplifying expressions

To **simplify** an expression, collect together like terms then add or subtract them according to their signs.

Worked example

1. Simplify $3x - 6y + 3x + 2y + 7$

    $(3x)(-6y)(+3x)(+2y)(+7)$

    $= (3x + 3x)(-6y + 2y)(+7)$ — Collect like terms.

    $= (6x)(-4y)(+7)$ — Simplify.

2. Simplify $2a^3 b - 3b^2 + a^3 b + 3b^2 + 5ab + 3a^2$

    $(2a^3b)(-3b^2) + (a^3b)(+3b^2)(+5ab)(+3a^2)$ — Collect like terms.

    $= (2a^3b + a^3b)(-3b^2 + 3b^2)(+5ab)(+3a^2)$ — These terms cancel each other out.

    $= (3a^3b)(+5ab)(+3a^2)$ — Simplify.

You can also simplify expressions that include powers or indices.

Worked example

1. Simplify:

    $\dfrac{12s^5t^2}{24st^7}$

    $\dfrac{12}{24} = \dfrac{1}{2}$ — Simplify the coefficients.

    $\dfrac{s^5}{s} = s^4$ — Simplify the s terms.

    $\dfrac{t^2}{t^7} = \dfrac{1}{t^5}$ or t^{-5} — Simplify the t terms.

    $\dfrac{12s^5t^2}{24st^7}$ — Put the terms together.

    $= \dfrac{s^4}{2t^5}$ or $\dfrac{1}{2}s^4 t^{-5}$

2. Simplify:

    $(3rs^2)^3$

    $3^3 = 27$ — Cube each of the terms.

    r^3

    $(s^2)^3 = s^{2 \times 3} = s^6$

    $27r^3 s^6$ — Put the terms together.

> **REVISION TIP**
> A coefficient is the number before a letter, e.g. the coefficient of $6m$ is 6.

Expanding and factorising single brackets

To **expand** a single bracket, multiply each term inside the bracket by the term in front of the bracket.

Worked example

Expand $3x(x + 7)$

Method 1: Using an area model

×	x	$+7$
$3x$	$3x^2$	$+21x$

$= 3x^2 + 21x$

Use the grid to multiply each term inside the bracket by the term outside the bracket.

Method 2: Drawing lines

$3x(x + 7)$

$= 3x^2 + 21x$

Use lines to help you multiply each term inside the bracket by the term outside the bracket.

Factorising is the opposite of expanding.

To factorise an expression:
- find the highest common factor (HCF) of the terms and put this outside the bracket
- work out what needs to go in the brackets.

Worked example

Fully factorise $12x^2y + 16xy^2$

HCF of 12 and 16 is 4.
HCF of x^2y and xy^2 is xy
so HCF $= 4xy$.

$4xy(\underline{} + \underline{})$

$\dfrac{12x^2y}{4xy} = 3x$ and $\dfrac{16xy^2}{4xy} = 4y$

$12x^2y + 16xy^2 = 4xy(3x + 4y)$

Find the HCF of the two terms. Put this outside the brackets.

Divide each of the terms by the HCF.

Add these terms inside the brackets.

Substitution

Substitution means replacing a variable with a number.

Worked example

Find the value of $abc + a^2$ when $a = -4$, $b = 5$ and $c = 2$.

$a \times b \times c + a^2$
$= (-4) \times (5) \times (2) + (-4)^2$
$= (-4) \times (5) \times (2) + 16$
$= -40 + 16$
$= -24$

Substitute all values.

Carry out the calculation in BIDMAS order.

REVISION TIP

Remember that squaring a negative gives a positive.

Knowledge

5 Algebra basics

Solving linear equations

To **solve** an equation, use **inverse operations** to move the unknown to one side on its own. Inverse operations are operations that are the opposite of each other. Examples include:

| $+$ and $-$ | $\times$ and $\div$ | x^2 and $\sqrt{x}$ |

Keep the equation balanced by doing the same operation to both sides.

> **REVISION TIP**
> In a linear equation, the highest power of the variable is always 1. $x + 5 = 7x$ is linear, but $x^2 + 2 = 3x$ is not.

For example, when solving $2x + 5 = 8 + x$:

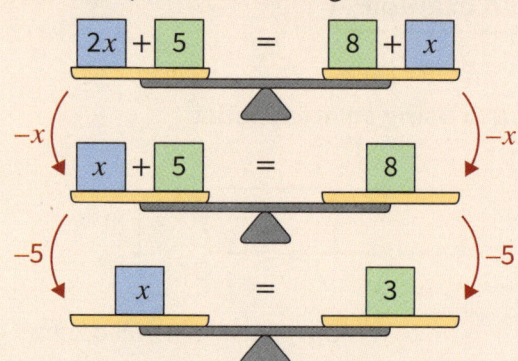

Worked example

1. Solve $\dfrac{3x}{2} + 4 = 12$

   $\dfrac{3x}{2} = 8$ — Subtract 4 from both sides.

   $3x = 16$ — Multiply by 2.

   $x = \dfrac{16}{3}$ — Divide by 3.

2. Solve $\dfrac{2(x+3)}{3} = x - 1$

   $\dfrac{\cancel{3} \times 2(x+3)}{\cancel{3}} = 3(x-1)$ — Get rid of the denominator by multiplying.

   $2x + 6 = 3x - 3$

   $2x + 6 - 2x = 3x - 3 - 2x$ — Expand the brackets.

   $6 = x - 3$

   $6 + 3 = x - 3 + 3$ — Use inverse operations to get x on only one side of the equation.

   $x = 9$

3. $a^3 \times \dfrac{a^{2n}}{a^4} = a^{-5}$

   Find the value of n. — Use inverse operations to get x on its own.

   $a^{-5} = a^3 \times \dfrac{a^{2n}}{a^4}$

   $= a^3 \times a^{2n} \times a^{-4}$ — Use laws of indices to simplify RHS.

   $= a^{3+2n-4}$

   $= a^{2n-1}$ — Equate the powers and solve for n.

   $-5 = 2n - 1$

   $2n = -4$

   $n = -2$

   > **REVISION TIP**
   > For this question, you need to use laws of indices to help you form a linear equation in terms of n.

A **formula** shows how one variable relates to another variable. For example:

- area = length × width
- $y = 2x + 5$
- distance = speed × time

Inverse operations can be used to rearrange formulae.

Worked example

1. The formula for the area of a circle is $A = \pi r^2$

   Rearrange the formula to give r in terms of A.

   $\dfrac{A}{\pi} = r^2$ — Divide both sides by π.

   $r = \sqrt{\dfrac{A}{\pi}}$ — Take the square root of both sides.

2. The area of a circle is 28.27 cm². Find the radius of this circle, correct to 1 d.p.

   $r = \sqrt{\dfrac{28.27}{\pi}}$ — Substitute the area into the formula and solve to find r.

   $= 3.0 \text{ cm}$

Solving linear equations

Worked example

The square and the triangle have the same perimeter.

13 cm, $(x+5)$ cm, x cm, x cm

Find the value of x.

perimeter of square $= 4x$ — Find expressions for the perimeters.
perimeter of triangle $= x + (x+5) + 13$
$= 2x + 18$ — Set the two expressions equal to each other.
$4x = 2x + 18$
$4x - 2x = 2x - 2x + 18$ — Simplify and solve for x.
$2x = 18$
$\frac{2x}{2} = \frac{18}{2}$ — Subtract $2x$.
$x = 9$ — Divide by 2.

REVISION TIP

If a question asks you to find the value of x but doesn't state an equation, use the information given to first form, and then solve, an equation.

An **identity** is an equation that is always true, no matter what the value of the variable is.

$2x \equiv x + x$

$2a \times b \equiv a \times 2b$

Sometimes, you will need to prove that an identity holds for any value of the variable.

The 'three-line' equals sign ($\equiv$) tells you that this is an identity. It is true no matter what the value of a.

You need to show that you can rearrange the left-hand side to make it the same as the right-hand side.

Worked example

1. Rearrange $\frac{1}{a} = \frac{1}{b} + \frac{1}{c}$ to make b the subject.

$\frac{abc}{a} = \frac{abc}{b} + \frac{abc}{c}$ — Multiply every term by the LCM of the denominators (abc) and simplify.
$bc = ac + ab$
$bc - ab = ac$ — Subtract ab so all terms involving b are on the same side of the equation.
$b(c - a) = ac$
$b = \frac{ac}{c - a}$ — Factorise.

Divide by $c - a$.

2. Show that $3(a + 2b) - 4a \equiv -a + 6b$

LHS $\equiv 3(a + 2b) - 4a$ — Expand the bracket.
LHS $\equiv 3a + 6b - 4a$
$\equiv -a + 6b$ — Collect like terms.
$\equiv$ RHS

Key terms

Make sure you can write a definition for these key terms.

equation   expand   expression   factorise   formula
identity   inequality   inverse operation   like terms
simplify   solve   substitute   term   variable

Retrieval

5 Algebra basics

Learn the answers to the questions below, then cover the answers column with a piece of paper and write as many as you can. Check and repeat.

	Questions	Answers
1	In algebra what is a term?	A single number or variable, or numbers and variables multiplied together e.g. 2, x, $5y$ are all examples of terms.
2	What are terms of the same type called?	Like terms.
3	What is the opposite of factorising?	Expanding.
4	What does it mean to expand a single bracket?	To multiply each term inside the bracket by the term in front of the bracket.
5	What is an expression?	A collection of letters and numbers which cannot be solved (no equals or inequality sign).
6	What is a coefficient?	The number before a letter, e.g. the coefficient of $6m$ is 6.
7	What is a variable?	A letter used to represent an unknown numerical value.
8	What does $2a$ mean?	2 multiplied by a.
9	What does a^n mean?	a raised to the power of n.
10	What does it mean to factorise fully?	To make sure that you have used the highest common factor, not just a common factor.
11	What is the difference between an equation and an identity?	An identity is true for all values of the variable.
12	What is a 'linear' equation?	An equation with no power higher than 1.

Previous questions

Now go back and use these questions to check your knowledge of previous topics.

	Questions	Answers
1	How do you convert a decimal into a percentage?	Multiply by 100 %.
2	What is the reciprocal of x?	$\frac{1}{x}$
3	What does HCF stand for?	Highest common factor.

Practice

Exam-style questions

5.1 Simplify

(a) $(2a)^3$ [2 marks]

(b) $\dfrac{6x^2y^{-3}}{18yx^{-1}}$ [2 marks]

(c) $\sqrt{x^4 y^6}$ [2 marks]

> **EXAM TIP**
> Convert the square root to index form.

5.2 Expand and simplify
$5(3 - x) - 4(6 - 3x)$. [2 marks]

5.3 Solve

(a) $\dfrac{5 - x}{2} = 12$ [2 marks]

(b) $\dfrac{2}{y} = 5$ [2 marks]

(c) $3 + p = 4p - 6$ [2 marks]

(d) $3(3 - 2p) = 4 - 11p$ [2 marks]

5.4 Solve
$5y - 2 = \dfrac{y - 13}{2}$ [3 marks]

5.5 Work out the value of x.

(a) $\dfrac{3^{-2} \times 3^8}{3^7} = 3x$ [3 marks]

(b) $2^5 \times 4^2 = 8x$ [3 marks]

5.6 Find the value of t, where $\dfrac{c^t \times c^{2t}}{c} = c^2$ [4 marks]

5.7 Rosalind thinks of a number. She multiplies her number by two, adds three and then writes the answer. Next, she multiplies her original number by three, subtracts four and writes the answer. Rosalind realises she has written the same answer both times. What is Rosalind's number? [3 marks]

> **EXAM TIP**
> Write and solve an equation using x as your unknown number.

Exam-style questions

5.8 Expand and simplify if possible.

(a) $(x + 2)(y + 1)$ [1 mark]

(b) $(2a - 5)(2 + a)$ [2 marks]

(c) $(a + b)^2$ [2 marks]

(d) $(2m + 3n)(3m - 2n)$ [2 marks]

(e) $(4x + 1)^2$ [2 marks]

5.9 Show that $(4x - 1)(2x + 5)(3x - 2)$ can be written in the form $ax^3 + bx^2 + cx + d$ where a, b, c and d are integers. [2 marks]

5.10 Factorise fully.

(a) $16x + 12xy$ [2 marks]

(b) $x^2y + y^2x$ [2 marks]

(c) $8p - 4p^2q + 6pq$ [2 marks]

> **EXAM TIP**
> When you're told to factorise 'fully', make sure you take out the HCF.

5.11 Make x the subject of the formulae.

(a) $2x^2 - 3 = y$ [2 marks]

(b) $a\sqrt{x} + b = c$ [2 marks]

5.12 The power, P watts, in a circuit is given by the formula

$$P = I^2 R$$

where I is the current in amps and R is the resistance in ohms.

Rearrange the formula to make I the subject. [2 marks]

5.13 Min-Jun is rearranging a formula to make x the subject. His steps are shown.

Explain where Min-Jun has gone wrong.

$$y = a^2 x - b$$
$$y + b = a^2 x$$
$$\sqrt{y + b} = ax$$
$$\frac{\sqrt{y + b}}{a} = x$$

[3 marks]

5.14 Make *r* the subject of the formula.

$$6d = \frac{n-7}{3r+1}$$ [3 marks]

5.15 Make *p* the subject of the formulae.

(a) $mp - q = ap$ [3 marks]

(b) $\frac{1}{p} + \frac{1}{r} = \frac{1}{t}$ [3 marks]

Questions referring to previous content

5.16 A six-character access code is needed for an online bank account. The first character must be a letter (A to Z). The remaining characters must be single digit (0 to 9). Numbers may be used more than once.

(a) Work out how many possible access codes there are. [2 marks]

(b) The rules for access codes are changed. Under the new rules, both the first and last characters of the code must be letters, with the remaining characters being single digit numbers. Under the new rules, no number or letter may be used more than once.

How many six-character access codes are possible now? [2 marks]

5.17 Write $0.78\dot{3}$ as a fraction. [3 marks]

Knowledge

6 Linear graphs

Graph terminology

A **coordinate grid** is a grid and a set of **axes** that you can plot points (called **coordinates**) and lines on.

The x-axis is a horizontal scale on the coordinate grid. The x-**intercept** is a point where a **graph** cuts the x-axis. In this graph, it is at $x = 3$.

The y-axis is the vertical scale. The y-intercept is a point where a graph cuts the y-axis. In this graph, it is at $y = 2$.

The **origin** is the point with coordinates $(0, 0)$.

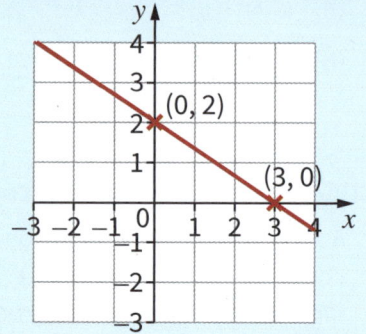

Linear graphs

A **linear** graph is a set of coordinates that make a straight line when plotted. This line can be horizontal, vertical or sloping.

$y = 3$ is a horizontal line

$y = 3$ goes through the y-axis at $(0, 3)$

$x = 3$ is a vertical line

$x = 3$ goes through the x-axis at $(3, 0)$

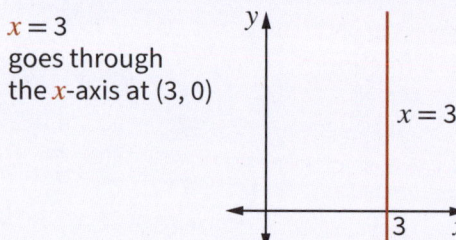

All lines of the form $y = a$ are horizontal for any value of a.

All lines of the form $x = a$ are vertical for any value of a.

Gradients

The **gradient** is a measure used to describe how steep a line graph is. It is how many units up (or down) the line goes for every one unit across.

To find the gradient of a line, first find two points on the line where you can read the exact coordinates. Then:

gradient = $\dfrac{\text{change in } y}{\text{change in } x}$

Which, in this example $= \dfrac{4}{2} = 2$

A line through the origin with gradient m has equation $y = mx$.

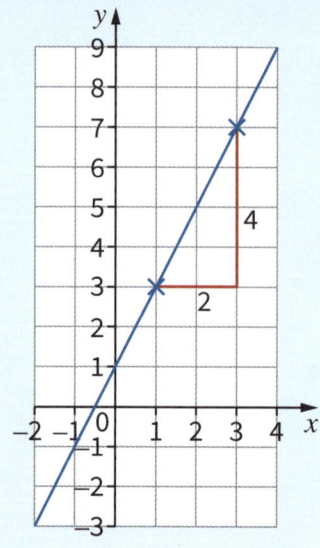

WATCH OUT
A gradient can be positive or negative.

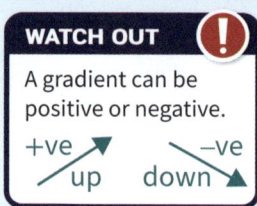

42   6 Linear graphs

6

Gradients

> **Worked example**
>
> 1. Calculate the gradient of this line.
>
>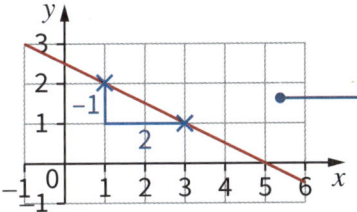
>
> Find two points where you can read the exact coordinates.
>
> gradient = $\dfrac{\text{change in } y}{\text{change in } x} = \dfrac{-1}{2} = -0.5$

Equation of a straight line

The **equation of a straight line** is a rule that is true for all points on the straight line. For example, $y = 2x + 5$.

Equations such as $y = x$ and $y = 2x + 5$ are sloping lines.

The general equation of a straight line is $y = mx + c$

- m → gradient
- c → intercept

A **table of values** is a way of writing down the values that satisfy a function or equation. For example:

for $y = 2x$:

x	0	1	2	3
y	0	2	4	6

for $y = 2x + 5$:

x	0	1	2	3
y	5	7	9	11

> **Worked example**
>
> 1. Fill in the table of values for the equation $y = 2x - 3$
>
> | x | −2 | −1 | 0 | 1 |
> |---|---|---|---|---|
> | y | | −5 | | |
>
> $y = 2 \times (-2) - 3 = -7$
> $y = 2 \times 0 - 3 = -3$
> $y = 2 \times 1 - 3 = -1$
>
> Substitute the values of x into the equation to find y
>
> | x | −2 | −1 | 0 | 1 |
> |---|---|---|---|---|
> | y | −7 | −5 | −3 | −1 |
>
> 2. Draw the graph of $y = 2x - 3$
>
>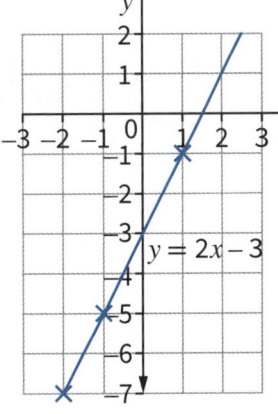
>
> Plot each of the points from the table $(-2, -7), (-1, -5), (0, -3), (1, -1)$
>
> Use a ruler to draw a straight line through all the points.
>
> **LINK**
> To remind yourself about rearranging equations, look back at Chapter 5.
>
> **EXAM TIP**
> When drawing graphs:
> - use a ruler
> - check the scales on both axes
> - the line must go all the way to the edge of the coordinate grid.

6 Knowledge 43

Knowledge

6 Linear graphs

Equation of a straight line

Worked example

1. A line has the equation $2y + 6 = 4x$

   Work out the gradient and the y-intercept of the line.

   $2y + 6 - 6 = 4x - 6$ — Rearrange the equation so it's in the form $y = mx + c$

   $2y = 4x - 6$

   $\frac{2y}{2} = \frac{4x - 6}{2}$

   $y = 2x - 3$

   The gradient (m) is 2 and the y-intercept (c) is -3 — Compare to $y = mx + c$

2. Find the equation of the line.

   gradient (m) = $\frac{\text{change in } y}{\text{change in } x} = \frac{3}{1} = 3$

   y-intercept (c) = 4

   $y = mx + c$

   $y = 3x + 4$

   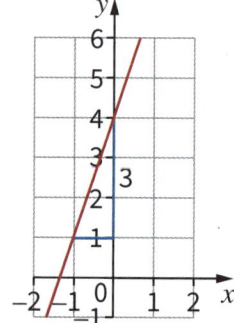

3. A line passes through the coordinates (1, 2) and (4, −7).

   Work out the equation of the line.

   gradient (m) = $\frac{\text{change in } y}{\text{change in } x} = \frac{-9}{3} = -3$

   $y = mx + c \Rightarrow y = -3x + c$

   $y = -3x + c$

   $2 = -3 \times 1 + c$ — To find c, substitute the coordinate pair (1, 2) into the equation (you could also have used the point (4, −7)).

   $2 = -3 + c$

   $c = 5$

   The equation is $y = -3x + 5$

Parallel and perpendicular lines

Two lines are **parallel** when they are always the same distance from each other.

A pair of parallel lines will have the same gradient.

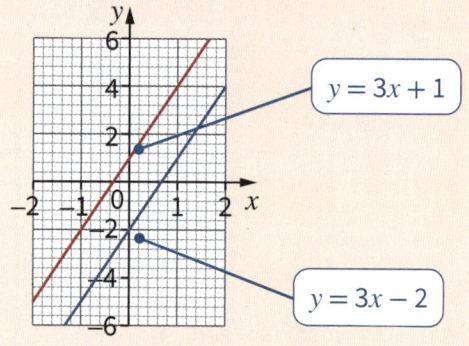

$y = 3x + 1$

$y = 3x - 2$

These lines are parallel. They both have a gradient of 3

If two lines are **perpendicular**, the gradient of one line is the negative reciprocal of the other line.

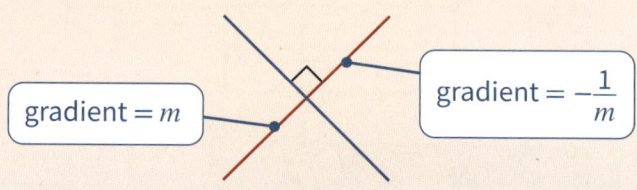

gradient = m

gradient = $-\frac{1}{m}$

Key terms Make sure you can write a definition for these key terms

axes coordinate coordinate grid
equation of a straight line gradient graph intercept
linear origin parallel perpendicular table of terms

Parallel and perpendicular lines

Worked example

A line, L, has equation $2y - x + 2 = 0$

Find the equation of the line that is perpendicular to L and passes through the point $(5, -4)$

$2y - x + 2 = 0$ — Rearrange into the form $y = mx + c$
$2y = x - 2$
$y = \frac{1}{2}x - 1$

gradient of $L = \frac{1}{2}$ — Identify gradient of L

negative reciprocal of $\frac{1}{2} = -2$ — Find gradient of the line perpendicular to L.

$y = -2x + c$ — Write the equation of the line.

$-4 = -2 \times 5 + c$ — Find the value of c for the perpendicular line by substituting the given coordinates into the equation.
$-4 = -10 + c$
$c = 6$

The equation is $y = -2x + 6$

Finding approximate values from a linear graph

Worked example

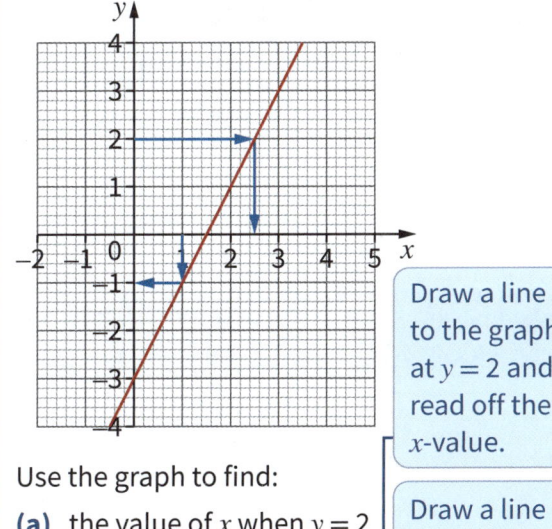

Use the graph to find:

(a) the value of x when $y = 2$

$x = 2.5$ — Draw a line to the graph at $y = 2$ and read off the x-value.

(b) the value of y when $x = 1$

$y = -1$ — Draw a line to the graph at $x = 1$ and read off the y-value.

Midpoints of lines

The midpoint of a line segment joining (x_1, y_1) and (x_2, y_2) is

$$\left(\frac{x_1 + x_2}{2}, \frac{y_1 + y_2}{2}\right)$$

Worked example

The straight line L_1 passes through the points with coordinates $(4, 7)$ and $(6, 3)$.

(a) Find the midpoint of the part of the line L_1 between $(4, 7)$ and $(6, 3)$.

midpoint $= \left(\frac{4+6}{2}, \frac{7+3}{2}\right)$ — Substitute the given coordinates into the formula.
$= (5, 5)$

(b) The straight line L_2 passes through the origin and has gradient $\frac{1}{2}$

The lines L_1 and L_2 intersect at the point P. Find the coordinates of P.

L_1: gradient $= \frac{3-7}{6-4} = -2$ — Write the equation of L_1 by finding the gradient and y-intercept using information in the question.
$y = -2x + c$

L_1 passes through $(6, 3)$.
$3 = -2 \times 6 + c$ — Substitute in $y = 3$ and $x = 6$.
$c = 15$
$L_1: y = -2x + 15$

L_2 has equation $y = \frac{1}{2}x$ — Write the equation of L_2
(y-intercept is 0 as it passes through the origin)

$y = -2x + 15$ ①
$y = \frac{1}{2}x$ ②

Eliminate y: $-2x + 15 = \frac{1}{2}x$ — To find where the lines intersect, write them as simultaneous equations and solve.
$-4x + 30 = x$
$30 = 5x$
$x = 6$

Sub $x = 6$ into ②: $y = \frac{1}{2} \times 6$
$= 3$

So the point of intersection is $(6, 3)$

6 Knowledge 45

Retrieval

6 Linear graphs

Learn the answers to the questions below, then cover the answers column with a piece of paper and write as many as you can. Check and repeat.

	Questions	Answers
1	In which directions do the x-axis and y-axis go?	The x-axis is horizontal and the y-axis is vertical.
2	What is a linear graph?	A linear graph is a straight line which can be horizontal, vertical or sloping.
3	What would the line $x = 4$ look like?	A vertical line passing through $x = 4$.
4	What would $y = -3$ look like?	A horizontal line passing through $y = -3$.
5	What is the general equation of a straight line?	$y = mx + c$
6	What is a first step to draw a linear graph when you know the equation?	Complete a table of values.
7	What does the gradient of the line measure?	How steep it is.
8	If a gradient is negative, what do you know about the line?	It goes downhill from left to right.
9	How do you calculate the gradient of a line?	$\dfrac{\text{change in } y}{\text{change in } x}$
10	What does m stand for in $y = mx + c$?	Gradient.
11	What does c stand for in $y = mx + c$?	y-intercept
12	What can you say about the gradients of parallel lines?	They are the same.
13	How would you find the gradient of a line in the following form: $ax + by = c$ (where a, b, c are integers)?	Rearrange into the form $y = mx + c$.

Previous questions

Now go back and use these questions to check your knowledge of previous topics.

	Questions	Answers
1	What does commutative mean?	A calculation can be done in any order.
2	When solving equations using the balance method, what do we do first if the unknown is on both sides?	Get the unknown on one side.
3	What is the opposite of factorising?	Expanding.

Practice

Exam-style questions

6.1 The straight line L_1 passes through the points $(-1, -3)$ and $(4, 2)$

The straight line L_2 passes through $(-2, -7)$ and has gradient 4.

The lines L_1 and L_2 intersect at the point T.

Find the coordinates of T. [4 marks]

6.2 Draw the graph of $x + y = 5$ on the grid.

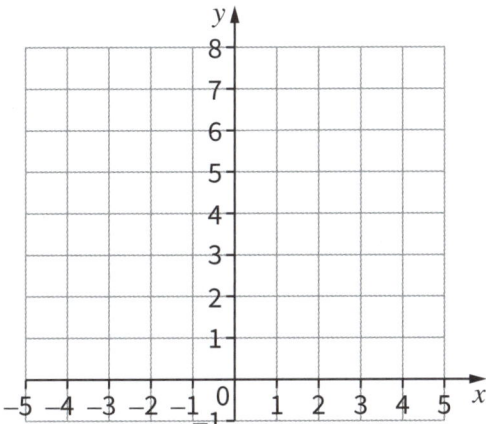

[3 marks]

6.3 On the grid, draw the graph of $y = 3 - 2x$

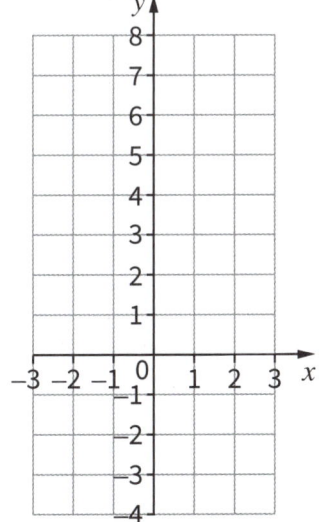

[3 marks]

Exam-style questions

6.4 Work out the gradient and y-intercept of each of these lines.

    (a) $y = 5x + 1$ [2 marks]

    (b) $y = 3 - 2x$ [2 marks]

    (c) $2y = x + 6$ [2 marks]

    (d) $y - x = 10$ [2 marks]

> **EXAM TIP**
> Rearrange the equations to make y the subject.

6.5 Write the equation of the line parallel to $y = 4x - 8$ that passes through the point (0, 5). [1 mark]

6.6 Work out the midpoint of the line segment that joins points (2, 6) and (4, 7). [2 marks]

6.7 Work out the equation of the line that passes through the two points.

    (a) (0, −1) and (2, 3) [3 marks]

    (b) (−3, 5) and (1, 1) [3 marks]

6.8 Match the equations of the lines on the left with the equations of perpendicular lines on the right.

$y = \frac{1}{2}x + 2$	$2y = 3x - 4$
$2x = 6 - 3y$	$y = x - 2$
$x + 2y - 1 = 0$	$y = -2x + 1$
$y + x = \frac{1}{2}$	$y - 2x = 0$

[3 marks]

6.9 The straight line L has equation $3x - 6y + 1 = 0$.

The point P has coordinates (4, −2).

Find an equation of the straight line that is perpendicular to L and passes through P. [4 marks]

6.10 The point A has coordinates (2, 1)

The point B has coordinates (p, q)

A line perpendicular to AB is given by the equation $2y + x = -1$

Find an expression for q in terms of p. [5 marks]

6.11 ABC is a right-angled triangle. L and M are straight lines.

L has equation $x = 3$.
$CO = \frac{1}{2} OB$
$CB = \frac{4}{3} BA$
Find an equation for M.

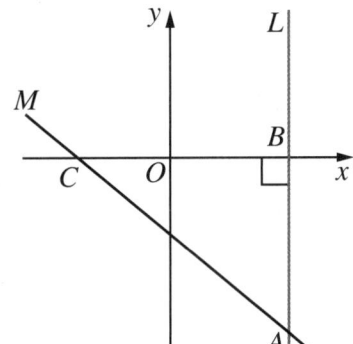

[5 marks]

Questions referring to previous content

6.12 Solve $3 + \frac{2(x+1)}{5} - x = 7 - (2 + x)$ [3 marks]

6.13 Find the value of $\frac{a - \sqrt{bc}}{c^2 - b}$ when $a = 10, b = -18, c = -2$ [4 marks]

Knowledge

7 Real-life graphs

Finding approximate values from a linear graph

Worked example

1. Find the value of x when $y = 2$.

   $x = 2.5$

2. Find the value of y when $x = 1$.

   $y = -1$

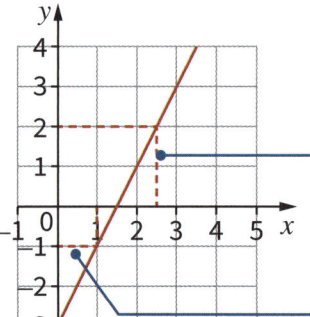

Draw a line to the graph at $y = 2$ and read off the x value.

Draw a line to the graph at $x = 1$ and read off the y value.

Real-life graphs

Real-life graphs are used to represent real-life situations and data. They can be straight lines or curves.

Worked example

This graph shows the cost of Peter's phone contract per month.

1. Explain why the graph starts at (0, 10).

   For 0 hours, Peter pays £10. This is a fixed cost before he makes any calls.

2. One month, Peter makes 40 hours of calls. Work out how much Peter's bill is.

   Look to see where the line is when the x value is 40 hours.

   £30

3. Determine how much a call costs per hour.

   You need to find the **gradient**.

   £5 ÷ 10 hours = £0.50 per hour of calls.

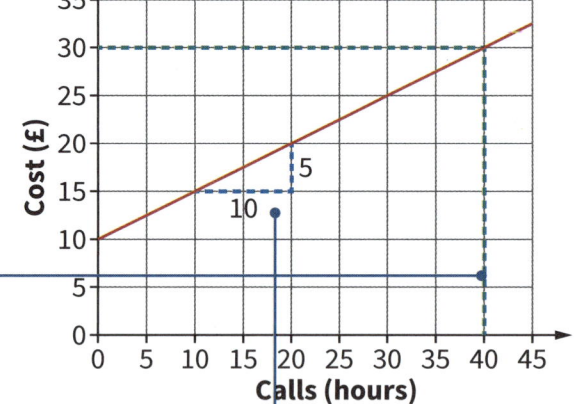

LINK

To remind yourself about gradients, look back at Chapter 6.

Formula

$$\text{gradient} = \frac{\text{change in } y}{\text{change in } x}$$

WATCH OUT

Always check what each unit on each scale represents.

50   7 Real-life graphs

7

Distance-time graphs

A distance-time graph shows how the distance changes as time passes.
On a distance-time graph, gradient = **speed**.

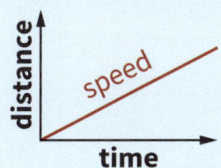

$$\text{gradient} = \frac{\text{distance}}{\text{time}}$$
$$= \textbf{speed}$$

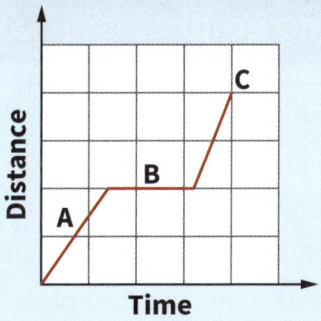

A. Straight sloped line means the speed is constant.
B. Horizontal section means time is passing but the distance isn't changing, so the speed is 0.
C. Steepest section, so the speed was the quickest here.

Worked example

The graph shows the height of a chilli plant at the end of each week.

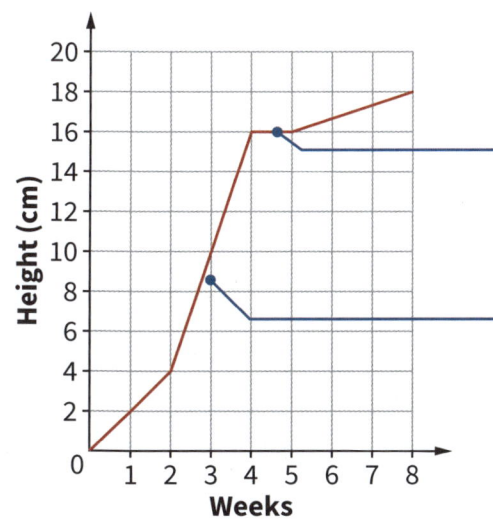

The line is flat between 4 and 5 weeks, which means the plant didn't increase in height in this time.

The line is steepest between 2 and 4 weeks, this means the height was increasing faster than at any other time.

1. How tall was the plant after three weeks?

   It was 10 cm tall.

2. In which week did the plant not grow?

   It did not grow between weeks 4 and 5.

3. When did the plant grow the fastest and what was its speed of growth during this time?

   The fastest increase was between 2 and 4 weeks. The speed of growth was $\frac{12}{2} = 6$ cm/week.

7 Knowledge

Knowledge

7 Real-life graphs

Distance–time graphs

Worked example

Min-Su drives a total of 14 km to work.
- She travels the first 6 km at a constant speed of 36 km/h.
- She then stops for 5 minutes at roadworks.
- It takes her 15 minutes to complete the rest of the journey at a constant speed.

Draw a distance–time graph of the journey.

Work out the time taken to travel first 6 km:

distance	time
36 km	1 hr
6 km	$\frac{1}{6}$ hr

($\div 6$ on both sides)

$\frac{1}{6}$ hr = $\frac{60}{6}$ = 10 min

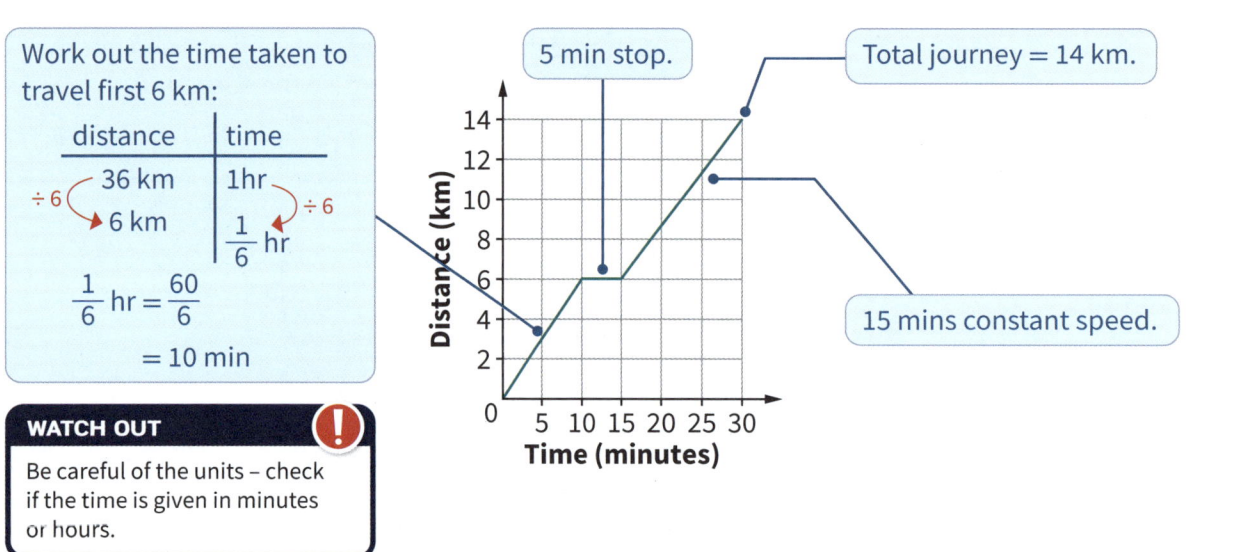

5 min stop.
Total journey = 14 km.
15 mins constant speed.

WATCH OUT
Be careful of the units – check if the time is given in minutes or hours.

Worked example

The distance–time graph shows how a snail moved one morning.
What was its speed in m/h between 09:00 and 09:30?

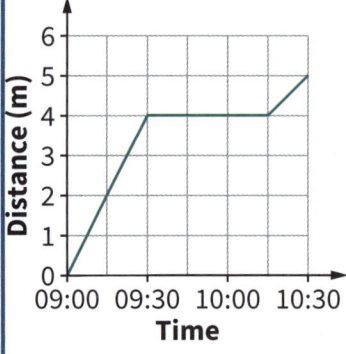

The snail moved 4 m in 0.5 hour.

distance	time
4 m	0.5 hr
8 m	1 hr

(×2 on both sides)

speed = 8 m/h — Use hours, not minutes here.

Key terms — Make sure you can write a definition for these key terms.

acceleration
real-life graph
speed

7 Real-life graphs

Speed–time graphs

A speed-time graph shows how the speed changes as time passes.
The gradient of a speed-time graph is equal to the **acceleration**.

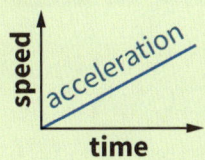

$$\text{gradient} = \frac{\text{speed}}{\text{time}}$$
$$= \textbf{acceleration}$$

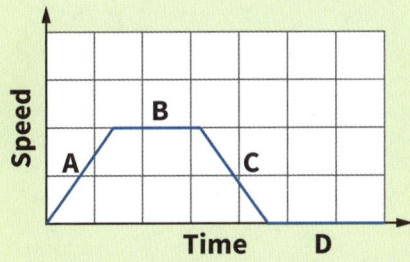

A. Straight sloped lines mean that acceleration is constant.

B. Horizontal section means time is passing but the speed isn't changing, so the speed is constant.

C. Negative gradient, which means deceleration or slowing down.

D. Speed = 0; they have stopped.

The area under the graph represents the distance travelled.

Rates of change

Worked example

These vases are filled with water at a constant rate. Sketch a graph for each to show how the depth of water varies over time.

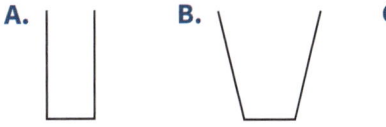

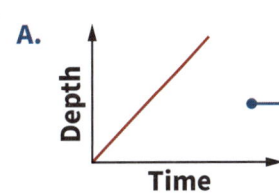

A. Depth of vase increases at constant rate so the graph is a straight line.

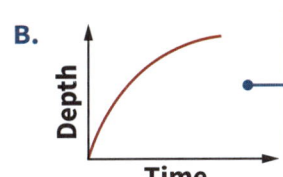

B. Rate of change of vase depth decreases as it gets wider, so the graph is a curve.

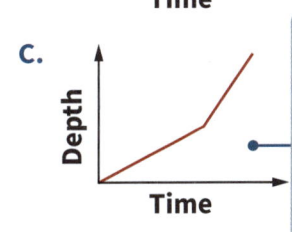

C. Graph has straight lines as the depth of the vase increases at a constant rate in both sections, but fills faster when it is narrower.

LINK

You will revise compound measures in Chapter 14.

Worked example

The graph shows the speed of a runner at the start of a 100 m race.

(a) Find their initial acceleration in m/s².
2.5 m/s²

(b) Find the distance covered by the runner in the first 4 seconds.
$\text{area} = \frac{1}{2} \times \text{base} \times \text{height}$
$= \frac{1}{2} \times 4 \times 10 = 20$ m.

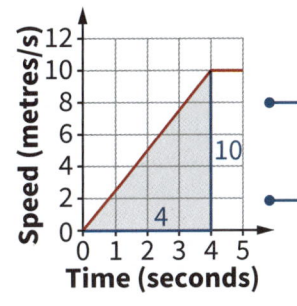

$\text{gradient} = \frac{\text{change in } y}{\text{change in } x}$
$= \frac{10}{4}$
$= 2.5$ m/s²

The distance travelled is the area under the graph between 0 and 4 seconds.

This is a triangle with a base of length 4 and a height of 10.

7 Knowledge

Retrieval

7 Real-life graphs

Learn the answers to the questions below, then cover the answers column with a piece of paper and write as many as you can. Check and repeat.

Questions | Answers

1. On a speed–time graph, what does the gradient represent? | Acceleration.

2. What does a straight line represent on a distance–time graph? | Constant speed.

3. On a distance–time graph, what does a horizontal section represent? | No movement; the speed is 0.

4. On a speed–time graph, what does a horizontal section represent? | Constant speed; the acceleration is 0.

5. How do you calculate speed from distance and time? | $\text{speed} = \dfrac{\text{distance}}{\text{time}}$

6. Why can't distance–time graphs have negative values? | You can't go back in time or have a negative distance.

7. How do you calculate distance from speed and time? | $\text{distance} = \text{speed} \times \text{time}$

8. What does a negative gradient mean on a speed–time graph? | The object is slowing down (decelerating).

Previous questions

Now go back and use these questions to check your knowledge of previous topics.

Questions | Answers

1. In a sale, a hat is reduced from \$34 to \$28. Find the percentage decrease. |
$$\% \text{ decrease} = \dfrac{\text{actual decrease}}{\text{original amount}} \times 100\%$$
$$= \dfrac{34 - 28}{34} \times 100\%$$
$$= 17.6\%$$

2. What is standard form used for? | To make it easier to understand very large or very small numbers.

3. Write $a^3 \div a^{-5}$ as a single power of a. |
$$a^3 \div a^{-5} = a^{3-(-5)}$$
$$= a^{3+5}$$
$$= a^8$$

Exam-style questions

7.1 Krystyna runs a race. Here is her velocity–time graph.

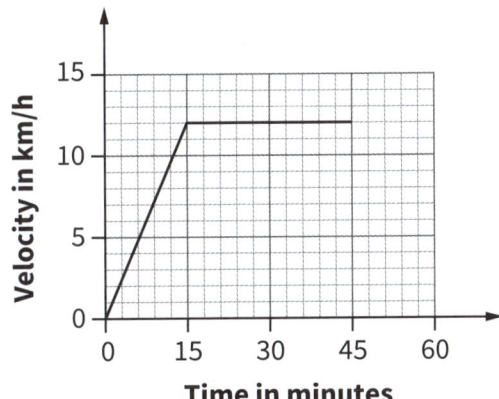

(a) After how many minutes does Krystyna finish the race? [1 mark]

(b) At what time is Krystyna's velocity 8 km/h? [1 mark]

(c) Describe how Krystyna's velocity changes through the race. [3 marks]

(d) Work out the total distance Krystyna travels during the race. [3 marks]

7.2 Kai cycles from his home.

Here is a distance–time graph showing part of Kai's journey.

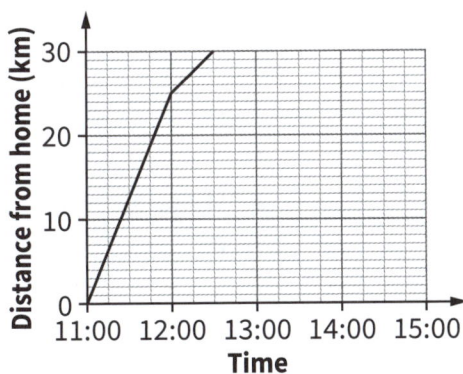

(a) For the first hour, Kai cycles at a speed of 25 km/h.
How far does he travel in the first hour and a half of his ride? [1 mark]

(b) At 12:30, Kai stops for lunch for 45 minutes before cycling back home at 20 km/h.

Complete the distance–time graph. [2 marks]

Exam-style questions

7.3 The graph shows the costs of taxi journey for different distances.

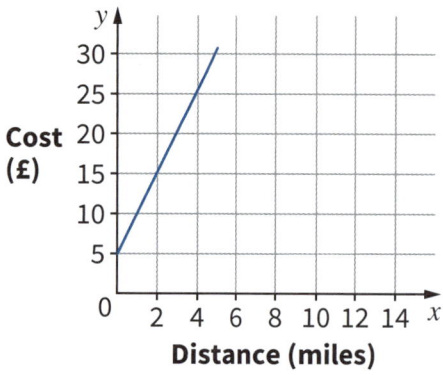

(a) Interpret the meaning of the y-intercept. [1 mark]

(b) Interpret the meaning of the gradient. [1 mark]

(c) How much would a 12-mile journey cost? [2 marks]

(d) State any assumptions that you have made in your answer to part (c). [1 mark]

7.4 The graph shows the height of garden plant A over a period of 6 weeks.

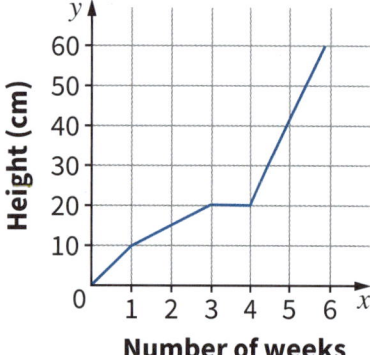

(a) In which week was there no growth? [1 mark]

(b) When was the growth fastest? [1 mark]

(c) Garden plant B grows at an average rate of 8 cm per week. Which garden plant grows slowest? Show your working. [2 marks]

Questions referring to previous content

7.5 The point with coordinates (−2, −1) lies on line L_1 with equation $y - 5 - 3x = 0$

The line L_2 also passes through (−2, −1) and is perpendicular to L_1.

Find the x-coordinate of the point on L_2 with y-coordinate −3. [4 marks]

7.6 Which two lines are perpendicular to each other? [2 marks]

$-2x + y = -2$ $5y + 2x = 10$ $\dfrac{x}{3} + y = -2$ $-2x + \dfrac{4}{5}y = -10$

Knowledge

8 Solving inequalities in 1 or 2 variables

Using inequalities

When two expressions are not equal, you can use an **inequality** to show which is bigger.

For example: $3x + 15 \geq 25$

> **REVISION TIP**
>
> Remember, the small end of the inequality sign points to the smaller value. For example, $5 < 9$ and $15 > 6$

When an inequality contains an **unknown** x, you can solve it to find the values of x that **satisfy** the inequality.

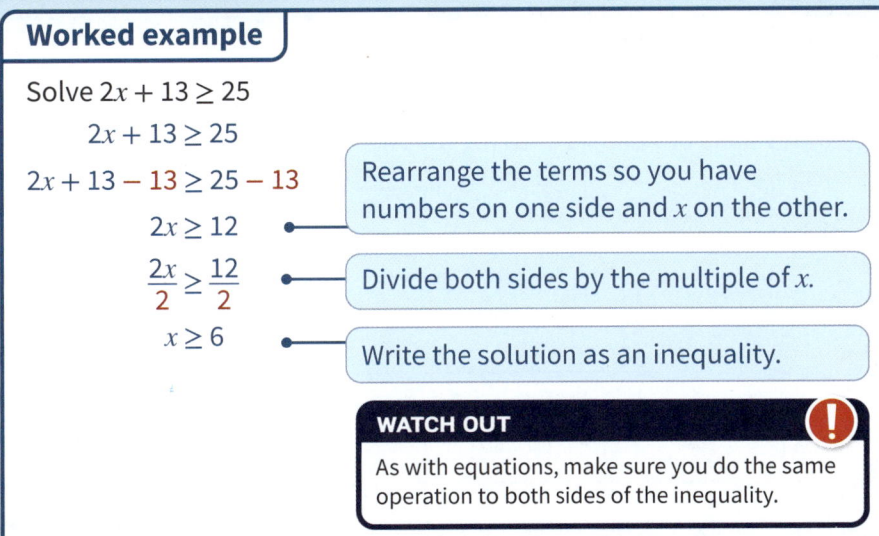

Worked example

Solve $2x + 13 \geq 25$

$2x + 13 \geq 25$

$2x + 13 - 13 \geq 25 - 13$ — Rearrange the terms so you have numbers on one side and x on the other.

$2x \geq 12$

$\dfrac{2x}{2} \geq \dfrac{12}{2}$ — Divide both sides by the multiple of x.

$x \geq 6$ — Write the solution as an inequality.

> **WATCH OUT**
>
> As with equations, make sure you do the same operation to both sides of the inequality.

If you multiply or divide both sides of an inequality by a negative number, the direction of the inequality changes.

Worked example

Solve $-2x + 13 \geq 25$

Method 1: Dividing by a negative and reversing the sign of the inequality.

$-2x + 13 \geq 25$ — Rearrange the terms so you have numbers on one side and x on the other.

$-2x \geq 25 - 13$

$-2x \geq 12$

$\dfrac{-2x}{-2} \leq \dfrac{12}{-2}$ — Divide both sides by the multiple of x.

$x \leq -6$ — Write the solution as an inequality.

> **WATCH OUT**
>
> Remember to change the direction of the inequality because it was divided by a negative number.

Method 2: Rearranging the equation so you only have to divide by a positive.

$-2x + 13 \geq 25$ — Take the x term to the other side of the equation to make it positive.

$13 \geq 25 + 2x$

$13 - 25 \geq 2x$

$-12 \geq 2x$ — Now take the number to the other side and simplify.

$-\dfrac{12}{2} \geq \dfrac{2x}{2}$ — Divide both sides by the multiple of x.

$-6 \geq x$

Write the solution as an inequality.

> **WATCH OUT**
>
> Inequality signs don't change when terms are divided by a positive.

You could be given two inequalities and asked to find values of x which satisfy both inequalities.

A **number line** can be used to represent the values of x that are included in the solution to the inequality.

A hollow circle shows that the number is not included (less/greater than).

A filled-in circle shows that the number is included (less/greater than or equal to).

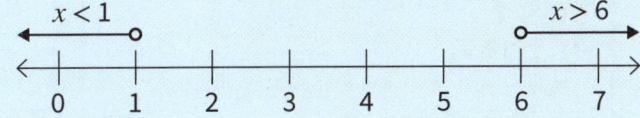

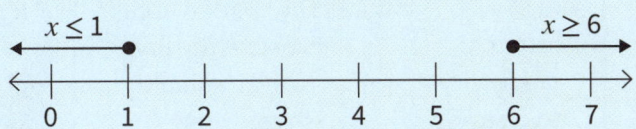

Knowledge

8 Solving inequalities in 1 or 2 variables

Using inequalities

Worked example

$2x + 1 > 7$ and $x - 4 \leq 5$

What values of x satisfy both inequalities? Show your answer on a number line.

$2x + 1 > 7$	$x - 4 \leq 5$
$2x > 6$	$x \leq 9$
$x > 3$	$3 < x \leq 9$

— Solve each inequality.

— Combine the inequalities. Take care to write the inequality sign the right way.

— Draw on a number line.

Representing inequalities on graphs

Solutions to a set of inequalities can be represented on a graph.

- **strict inequalities** (< and >) are represented with a dotted line.
- **non-strict inequalities** (≤ and ≥) are represented with a solid line.

Worked example

Write an inequality to satisfy the shaded region.

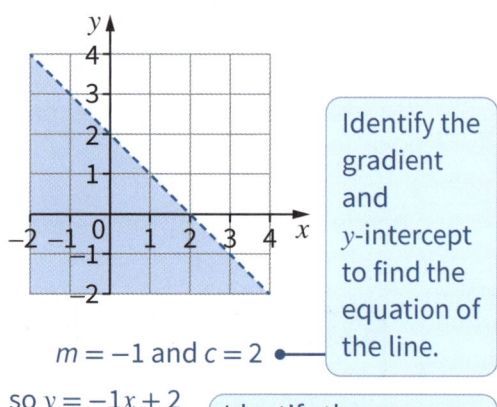

$m = -1$ and $c = 2$

so $y = -1x + 2$
$= 2 - x$

$y < 2 - x$

$0 < 2 - 0$ ✓

Substitute $(0,0)$ into the inequality to check it.

Identify the gradient and y-intercept to find the equation of the line.

Identify the inequality:
- A dotted line means it is not included in the inequality.
- The shaded area is below the line, so the inequality is 'less than'.

Worked example

Shade the region that satisfies the inequalities $y > 4 - 2x$, $x \leq 7$, and $y \leq 3$.

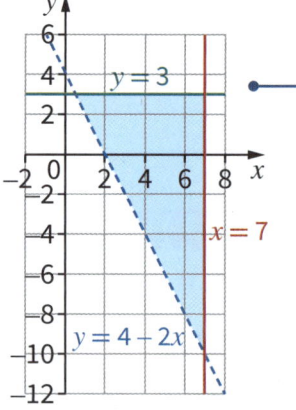

Draw the lines $y = 4 - 2x$, $x = 7$ and $y = 3$.

Use a dotted line for $y = 4 - 2x$ because it is a strict inequality.

Consider each inequality in turn:
- $x \leq 7$, so you need x-values smaller than or equal to 7.
- $y \leq 3$, so you need y-values smaller than or equal to 3.
- For $y > 4 - 2x$, the inequality is 'greater than' so the shaded area is above the line.

Shade the area that satisfies all the inequalities.

Key terms — Make sure you can write a definition for these key terms

expression   inequality
non-strict inequality
strict inequality   number line
satisfy   solve   unknown

8 Solving inequalities in 1 or 2 variables

Retrieval

8 Solving inequalities in 1 or 2 variables

Learn the answers to the questions below, then cover the answers column with a piece of paper and write as many as you can. Check and repeat.

Questions | Answers

#	Question	Answer
1	What is an inequality?	A relationship between two expressions or values that are not equal to each other.
2	How do you represent an inequality on a number line where the value is **not** included?	Draw a hollow circle.
3	What should you do when you multiply or divide an inequality by a negative number?	Change the direction of the inequality symbol.
4	What does a filled-in circle mean when an inequality is represented on a number line?	The number at the circle is included in the inequality.
5	What does the $a > b$ mean?	a is greater than b.
6	What does the $a \leq b$ mean?	a is less than or equal to b.
7	How do you solve two inequalities that require you to show the solutions on a number line?	Solve each inequality separately then combine the solutions.
8	On a graph, would you use a dotted or solid line to represent the inequality $2x > 5$?	Dotted, as the points on the line do not satisfy the inequality.
9	On a graph, would you use a dotted or solid line to represent the inequality $6 \geq 5x$?	Solid, as the points on the line satisfy the inequality.

Previous questions

Now go back and use these questions to check your knowledge of previous topics.

Questions | Answers

#	Question	Answer
1	How do you convert a decimal into a percentage?	Multiply by 100%.
2	What does truncate mean?	Remove digits without rounding.
3	What does it mean to expand a single bracket?	To multiply each term inside the bracket by the term in front of the bracket.
4	When a number is written in standard form, what form does it take?	$a \times 10^n$, where $1 \leq a < 10$ and n is an integer
5	How do you multiply fractions?	Multiply the numerators and multiply the denominators.

8 Retrieval

Practice

Exam-style questions

8.1 Solve these inequalities and show the solutions on number lines.

(a) $3x + 5 > 2$

[3 marks]

(b) $20 - 5x \geq 0$

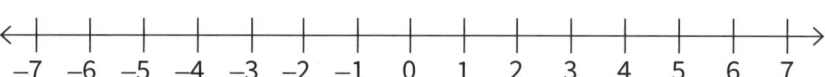

[3 marks]

EXAM TIP

Remember to include the inequality symbol in your final answers.

8.2 (a) Solve $-6 < x - 4 \leq -1$ [2 marks]

(b) Show the solution on the number line.

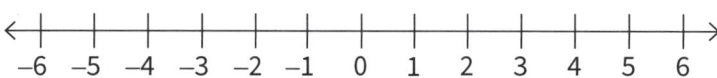

[1 mark]

8.3 Find the set of values of x that satisfies both inequalities.

$4x + 1 \geq 3$

$5 - x > 3$ [3 marks]

8.4 Write the inequality that describes each of these shaded regions.

(a) (b)

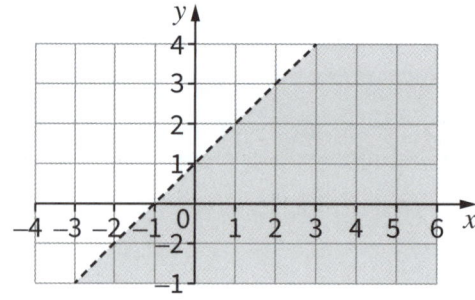

[1 mark]   [1 mark]

8.5 Draw axes with $-4 \leq x \leq 4$ and $-4 \leq y \leq 4$.
Shade the region that satisfies these inequalities.

$y < 2x \quad y \geq -2 \quad x + y \geq 1$

Label the region R.

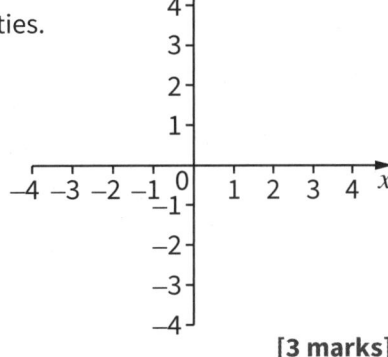

[3 marks]

8.6 **(a)** Write the three inequalities that define the shaded region. **[3 marks]**

(b) Write the points in the region with integer coordinates. **[1 mark]**

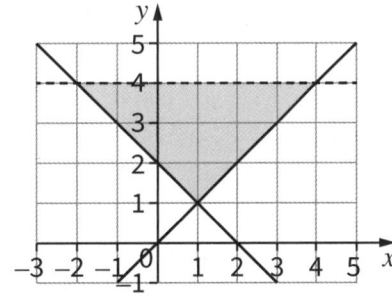

8.7 If I halve my number and subtract 4, I get more than if I double my number and add 3.

Form an inequality and solve it to find the range of possible values for my number. **[4 marks]**

8.8 The width of a rectangle is x cm. The length is 4 cm longer than the width.

The perimeter of the rectangle is less than 20 cm.

Form an inequality and solve it to find to find the range of possible values for x. **[4 marks]**

Questions referring to previous content

8.9 Draw a distance-time graph to represent Alex's journey to school. **[5 marks]**

She travels for 25 minutes at 24 km/h.

She then gets stuck in traffic for 9 minutes.

After the traffic clears, she drives the remaining 15 km in 11 minutes.

Use your graph to find Alex's average speed over the whole journey to school.

8.10 With regards to speed-time graphs, which one of the following is the only true statement? **[1 mark]**

A A line segment with zero gradient means the object is at rest.

B A line segment with negative gradient means the object is moving backwards.

C An upwards-sloping line segment represents constant acceleration.

D The maximum point on the graph represents maximum acceleration.

Knowledge

9 Factorising quadratic expressions and solving quadratic equations

Quadratic expressions

A **quadratic expression** is an expression where the highest power of the variable is 2.

They are of the form $ax^2 + bx + c$, where a, b and c are numbers. a is the **coefficient** of the x^2 term and b is the coefficient of the x term.

For example, $x^2 + 5x - 7$ is a quadratic expression, where $a = 1$, $b = 5$, and $c = -7$.

If the coefficient b is 0, the expression is: $ax^2 - c$

If the coefficient c is 0, the expression is: $ax^2 + bx$

Expanding double brackets

To **expand** (multiply out) a pair of brackets, multiply each term in the first set of brackets by each term in the second pair of brackets.

$(x + 5)(2x - 7) = 2x^2 - 7x + 10x - 35$
$\qquad\qquad\qquad = 2x^2 + 3x - 35$ ← Collect like terms.

If you prefer, you can use a multiplication grid.

×	x	5
$2x$	$2x^2$	$10x$
-7	$-7x$	-35

Difference of two squares

The **difference of two squares** is an expression, such as $a^2 - b^2$, which contains one number squared minus another number squared. For example, $x^2 - 9$ is the difference of two squares.

Formula

The difference of two squares can be factorised:
$a^2 - b^2 = (a + b)(a - b)$

When you multiply out the brackets, the middle terms $-ab$ and $+ab$ cancel out and leave just $a^2 - b^2$

×	a	$+b$
a	a^2	$+ab$
$-b$	$-ab$	$-b^2$

$a^2 + ab - ab - b^2$
$= a^2 - b^2$

Worked example

Factorise $4x^2 - 9$

$4x^2 = (2x)^2$ and $9 = 3^2$,
so, $4x^2 - 9 = (2x)^2 - 3^2$
$\qquad\qquad\;\; = (2x - 3)(2x + 3)$
$\qquad\qquad\;\; = 4x^2 - 6x + 6x - 9$ ← Check by multiplying out.
$\qquad\qquad\;\; = 4x^2 - 9$

Solving quadratic equations by factorising

You can solve some quadratic equations by **factorising**.

Worked example

Solve $x^2 + 6x = 7$ ← Rearrange into the form $ax^2 + bx + c = 0$

$x^2 + 6x - 7 = 0$
$x^2 - x + 7x - 7 = 0$ ← Factorise into double brackets by finding two numbers that add to give 6 and multiply to give -7
$x(x - 1) + 7(x - 1) = 0$
$(x + 7)(x - 1) = 0$

$x + 7 = 0$ ← Set each factor equal to zero.
$x - 1 = 0$
$x = -7$
$x = 1$

Numbers are 7 and -1:
$7 + (-1) = 6$
$7 \times (-1) = -7$

Solve each new equation and write the values of x.

Factorising quadratic equations

Factorising is the opposite of expanding.

Some expressions in the form $ax^2 + bx + c$ can be factorised into a double set of brackets.
For example: $x^2 + 6x + 5 = (x + 5)(x + 1)$

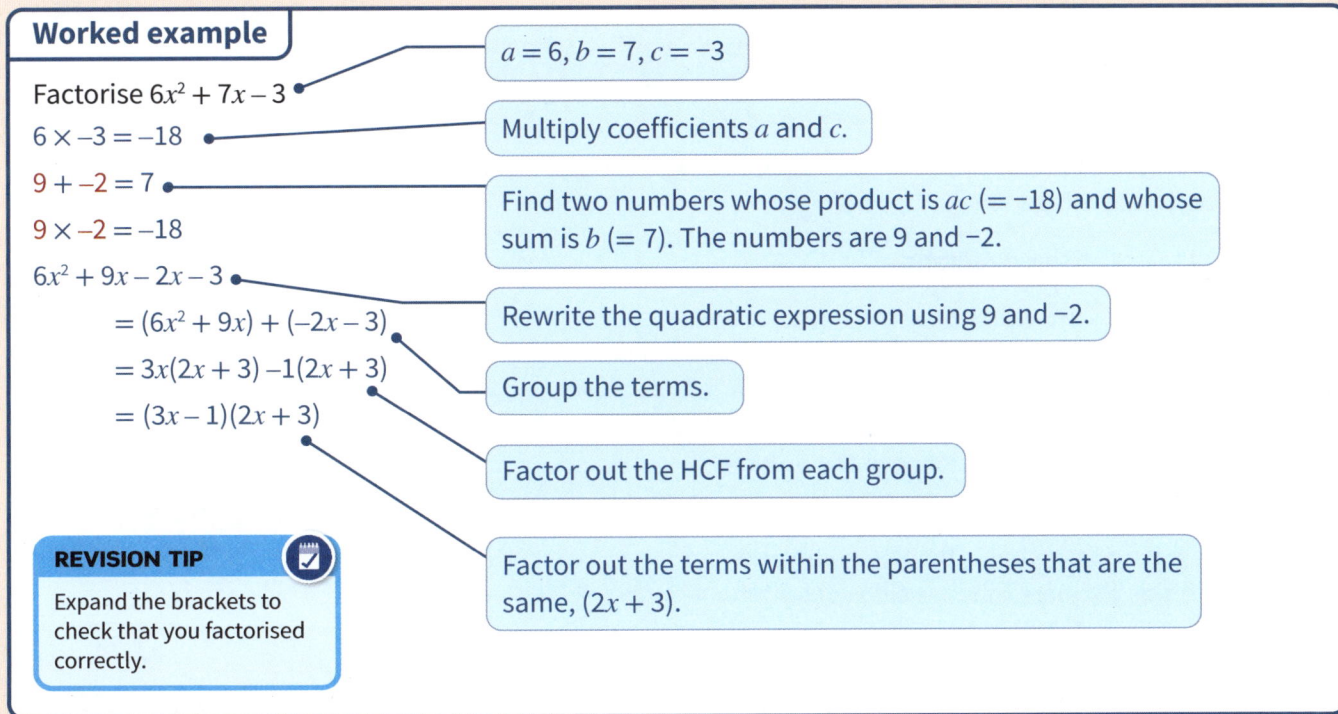

Worked example

Factorise $6x^2 + 7x - 3$

$6 \times -3 = -18$

$9 + -2 = 7$

$9 \times -2 = -18$

$6x^2 + 9x - 2x - 3$

$= (6x^2 + 9x) + (-2x - 3)$

$= 3x(2x + 3) - 1(2x + 3)$

$= (3x - 1)(2x + 3)$

- $a = 6, b = 7, c = -3$
- Multiply coefficients a and c.
- Find two numbers whose product is ac ($= -18$) and whose sum is b ($= 7$). The numbers are 9 and -2.
- Rewrite the quadratic expression using 9 and -2.
- Group the terms.
- Factor out the HCF from each group.
- Factor out the terms within the parentheses that are the same, $(2x + 3)$.

REVISION TIP

Expand the brackets to check that you factorised correctly.

Quadratic expressions that have two terms, and a common factor, e.g. $ax^2 + bx$ can be factorised into single brackets. For example: $x^2 + 3x = x(x + 3)$

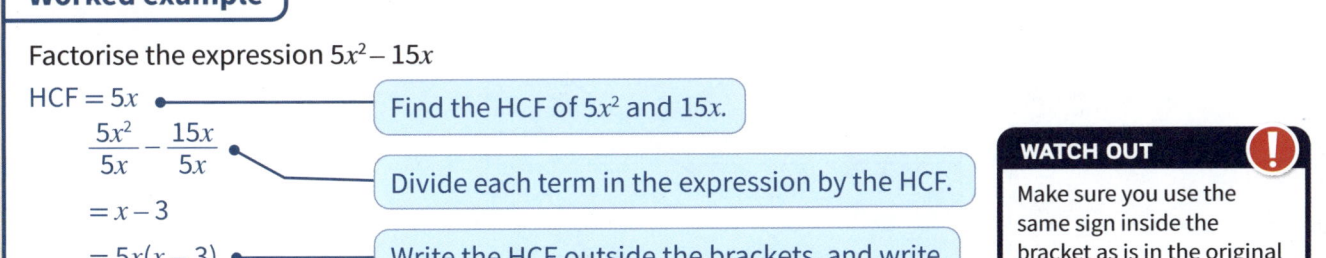

Worked example

Factorise the expression $5x^2 - 15x$

HCF $= 5x$

$\dfrac{5x^2}{5x} - \dfrac{15x}{5x}$

$= x - 3$

$= 5x(x - 3)$

- Find the HCF of $5x^2$ and $15x$.
- Divide each term in the expression by the HCF.
- Write the HCF outside the brackets, and write the divided terms inside the bracket.

WATCH OUT

Make sure you use the same sign inside the bracket as is in the original expression.

Knowledge

9 Factorising quadratic expressions and solving quadratic equations

Solving quadratic equations by completing the square

Quadratic equations that can't be solved by factorising can be solved by 'completing the square'. The factorised equation takes the form $a(x + p)^2 + q$, where a, p and q are numbers.

Worked example

By completing the square, write $x^2 + 6x + 17$ in the form $(x + p)^2 + q$, where p and q are integers.

- You can't factorise this quadratic.
- Write the expression without the constant term.

- Halve the coefficient of x to give you the constant term in the squared bracket. Replace the numerical term with $-c$ to retain the balance of the original equation.

- Since we have a term that is being squared, we can picture this as a square that has sides of length $x + 3$.

REVISION TIP

We can also find c by expanding the brackets.

$(x + 3)^2$
$= x^2 + 3x + 3x + 9$
$= x^2 + 6x + 9$

Remember to use the value of $-c$ to complete the square.

$x^2 + 6x = (x + 3)^2 - 9$

$x^2 + 6x$
$= (x + 3)^2 - c$

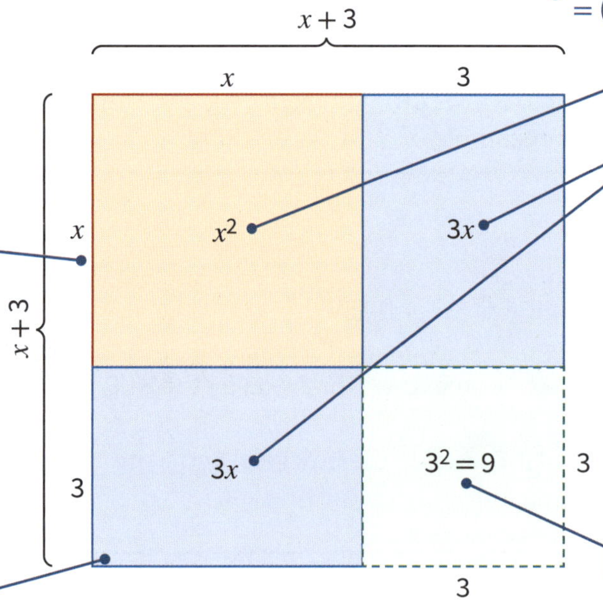

- This square has an area of x^2.
- These two rectangles have a combined area of $6x$.

- Each side of the square has length $x + 3$

- The large square has area $(x + 3)^2$

- The missing area is $3^2 = 9$

- The square of area 9 is missing from the expression, so to complete the square, subtract this from the factorised expression.

$x^2 + 6x = (x + 3)^2 - 9$
$x^2 + 6x + 17 = (x + 3)^2 - 9 + 17$
$= (x + 3)^2 + 8$

- Add the constant term from the original expression into the factorised expression.

- Simplify.

Solving quadratic equations by completing the square

Quadratic equations may need to be rearranged before you can solve them.

Worked example

Solve $2x^2 = 8x + 14$. Give your answer in the form $a \pm b\sqrt{c}$, where a, b and c are integers.

$2x^2 = 8x + 14$ — Divide through by 2 so the coefficient of x^2 is 1

$x^2 = 4x + 7$

$x^2 - 4x - 7 = 0$ — Write as a quadratic expression equal to 0

$x^2 - 4x = (x - 2)^2 - k$ for some value of k — Complete the square.

$x^2 - 4x = x^2 - 4x + 4 - k$

so $k = 4$

$x^2 - 4x = (x - 2)^2 - 4$

So $x^2 - 4x - 7 = 0$ becomes:

$(x - 2)^2 - 4 - 7 = 0$

$(x - 2)^2 - 11 = 0$

$(x - 2)^2 = 11$

$x - 2 = \pm\sqrt{11}$ — Take the square root of both sides and rearrange to make x the subject.

$x = 2 \pm \sqrt{11}$

Solving quadratic equations using the quadratic formula

The method for completing the square for the general quadratic equation $ax^2 + bx + c = 0$, gives the quadratic formula:

$$x = \frac{-b \pm \sqrt{b^2 - 4ac}}{2a}$$

Worked example

Solve $5x^2 + 7x - 3 = 0$

$a = 5$, $b = 7$, $c = -3$ — Identify the values of a, b and c.

$x = \dfrac{-7 \pm \sqrt{7^2 - 4 \times 5 \times (-3)}}{2 \times 5}$ — Substitute these values into the quadratic formula.

$x = \dfrac{-7 \pm \sqrt{109}}{10}$

$x = \dfrac{-7 + \sqrt{109}}{10} \quad x = \dfrac{-7 - \sqrt{109}}{10}$ — Solve to find the two possible values for x.

$x = 0.344 \quad x = -1.74$ (to 3 s.f.)

Key terms Make sure you can write a definition for these key terms.

coefficient   difference of two squares   expand
factor   factorise   quadratic expression   solve

Retrieval

9 Factorising quadratic expressions and solving quadratic equations

Learn the answers to the questions below, then cover the answers column with a piece of paper and write as many as you can. Check and repeat.

	Questions	Answers
1	What is a quadratic equation?	An equation where the highest power of the variable is 2.
2	What should be your first step if you want to solve $x^2 + x = 12$?	Subtract 12 from both sides so the equation is equal to 0.
3	If you factorise $x^2 + bx + c$ you need to find two numbers which add to give which letter?	b
4	If you factorise $x^2 + bx + c$ you need to find two numbers which multiply to give which letter?	c
5	What is an expression in the form $a^2 - b^2$ called?	The difference of two squares.
6	What does $a^2 - b^2$ factorise to?	$(a-b)(a+b)$
7	What is the general form of a quadratic expression?	$ax^2 + bx + c$
8	How do you solve a quadratic expression that can be factorised?	Rearrange the equation such that $ax^2 + bx + c = 0$ then factorise.
9	When might you use the quadratic equation?	To solve a quadratic equation that cannot be factorised.
10	What form do you write a quadratic expression in when you complete the square?	$a(x+p)^2 + q$

Previous questions

Now go back and use these questions to check your knowledge of previous topics.

	Questions	Answers
1	What does the letter I in BIDMAS stand for?	Index or indices.
2	What does $5a$ mean?	5 multiplied by a.
3	What is the opposite of expanding?	Factorising.
4	What does the gradient of a line tell you about the line?	How steep the line is.
5	What is rationalising the denominator?	Rewriting a fraction to remove any surds from the denominator.

Practice

Exam-style questions

9.1 A cube has side length $(x + 1)$ cm.

Find an expression for the volume of the cube in the form $ax^3 + bx^2 + cx + d$. [3 marks]

9.2 Factorise

(a) $p^2 - 9p - 36$ [2 marks]

(b) $x^2 + 9x - 10$ [2 marks]

9.3 A rectangle has area $x^2 + 6x - 27$

Find expressions for the lengths of the two sides of the rectangle.

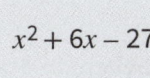

$x^2 + 6x - 27$

[2 marks]

9.4 Factorise fully

(a) $y^2 - 81$ [1 mark]

(b) $4b^2 - 1$ [1 mark]

(c) $28 + 4y^2$ [1 mark]

9.5 (a) Factorise $p^2 - q^2$ [1 mark]

(b) Hence simplify fully $(3 - x)^2 - (3x + 3)^2$ [3 marks]

9.6 Factorise

(a) $6x^2 + 5x - 4$ [2 marks]

(b) $12x^2 + 20x + 3$ [2 marks]

9.7 A square has area $9x^2 + 30x + 25$. Find the perimeter of the square in terms of x. [2 marks]

Exam-style questions

9.8 Solve

    (a) $x^2 - 7x - 8 = 0$ [3 marks]

    (b) $x^2 + 5x = 0$ [3 marks]

    (c) $x^2 - 1 = 0$ [3 marks]

9.9 Solve

    (a) $x^2 - 12x = -35$ [4 marks]

    (b) $x^2 + 3x = 10$ [4 marks]

EXAM TIP
Before solving, make sure that one side of the equation is zero.

9.10 Here is a rectangle. The area of the rectangle is 5 cm².

    (a)   Show that $2x^2 - 5x - 3 = 0$ [3 marks]

    (b)   Find the length of the shorter side of the rectangle.   [3 marks]

Rectangle: 5 cm², $(x - 2)$ cm, $(2x - 1)$ cm

9.11 Solve these equations. Give your answers to 3 significant figures where appropriate.

    (a) $3x^2 - 7x - 10 = 0$ [3 marks]

    (b) $5x^2 + 4x - 20 = 0$ [3 marks]

    (c) $25 + 10x - 2x^2 = 0$ [3 marks]

9.12 Solve these equations.

Give your answers in the form $a \pm b\sqrt{c}$, where a, b and c are integers.

    (a) $x^2 - 10x + 7 = 0$ [3 marks]

    (b) $x^2 + 2x - 11 = 0$ [3 marks]

EXAM TIP
You can either use the quadratic formula or completing the square to solve these equations.

9.13 Pavel says that $2x^2 + 7x + 6$ factorises to $(2x + 6)(x + 1)$, since the sum of 6 and 1 is 7 and the product of 6 and 1 is 6. Pavel is wrong. Explain why.   [2 marks]

9.14 Write these expressions in the form $(x + p)^2 + q$

    (a) $x^2 + 6x + 5$ [2 marks]

    (b) $x^2 - 4x + 10$ [2 marks]

    (c) $x^2 - 10x$ [2 marks]

    (d) $x^2 + x + 1$ [2 marks]

9.15 (a) Complete the square for the expression $x^2 - 2x - 11$ [2 marks]

(b) Hence, solve the equation $x^2 - 2x - 11 = 0$

Give your answer in the form $a \pm b\sqrt{3}$, where a and b are integers. [2 marks]

9.16 Write these expressions in the form $a(x + p)^2 + q$

(a) $2x^2 + 8x + 7$ [3 marks]

(b) $12 + 10x - x^2$ [3 marks]

9.17 (a) Complete the table of values for $y = x^2 + 5x - 2$

x	−6	−5	−4	−3	−2	−1	0	1
y		−2			−8			4

[2 marks]

(b) On the grid, draw the graph of $y = x^2 + 5x - 2$ for values of x from −6 to 1

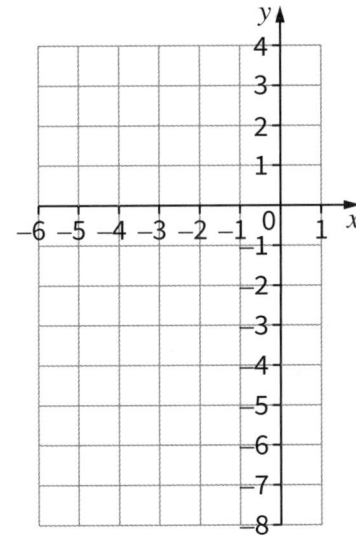

[2 marks]

(c) Use the graph to estimate a solution to $x^2 + 5x - 2 = 0$ [1 mark]

Questions referring to previous content

9.18 Solve $2(x + 1) - 4 \leq 3\left(x - \dfrac{4}{3}\right) < 3 + 2x$ [3 marks]

9.19 On a coordinate grid, sketch the following inequalities and describe the shape that they bound. [4 marks]

$x + y > 4 \qquad 3y - 7x > -28 \qquad 7y - 3x \leq 28$

Knowledge

10 Quadratic graphs, iterations, solving quadratic inequalities

Quadratic graphs

A **quadratic graph** is one with the general equation:

$y = ax^2 + bx + c$

where a, b, and c are numbers.

Roots are the solutions of $ax^2 + bx + c = 0$. These are the x-intercepts of the graph.

The maximum or minimum point of the graph is called the **turning point**.

All quadratic graphs have a **line of symmetry** through the turning point.

Quadratic graphs may have 2, 1 or 0 intercepts.

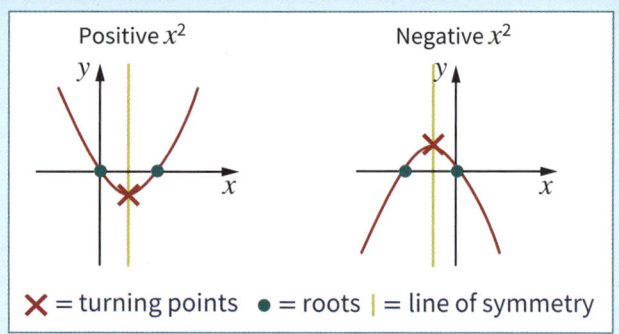

✗ = turning points   ● = roots | = line of symmetry

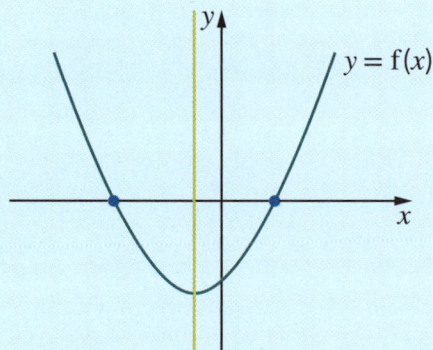

2 x-intercepts, so f(x) has 2 real roots.

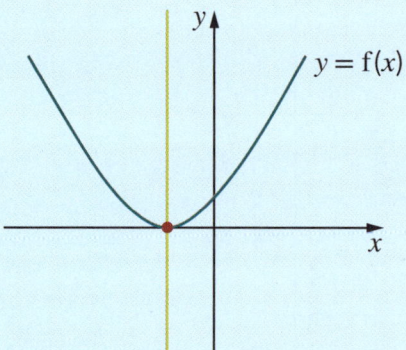

1 x-intercept, so f(x) has 1 real root.

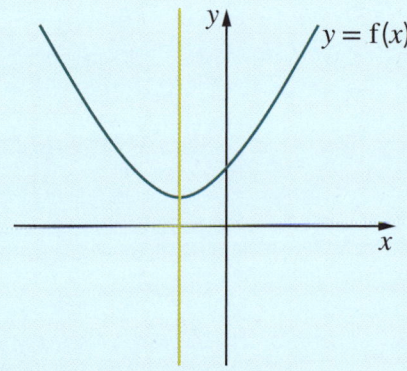

0 x-intercepts, so f(x) has 0 real roots.

Roots, intercepts, and turning points can be found using a graph.

Worked example

Here is the graph $y = x^2 + 2x - 1$

1. Write down the coordinates of the turning point of $y = x^2 + 2x - 1$

   $(-1, -2)$

2. Write down estimates for the roots of $x^2 + 2x - 1 = 0$

   Approximately, $x = 2.4$ and $x = 0.4$

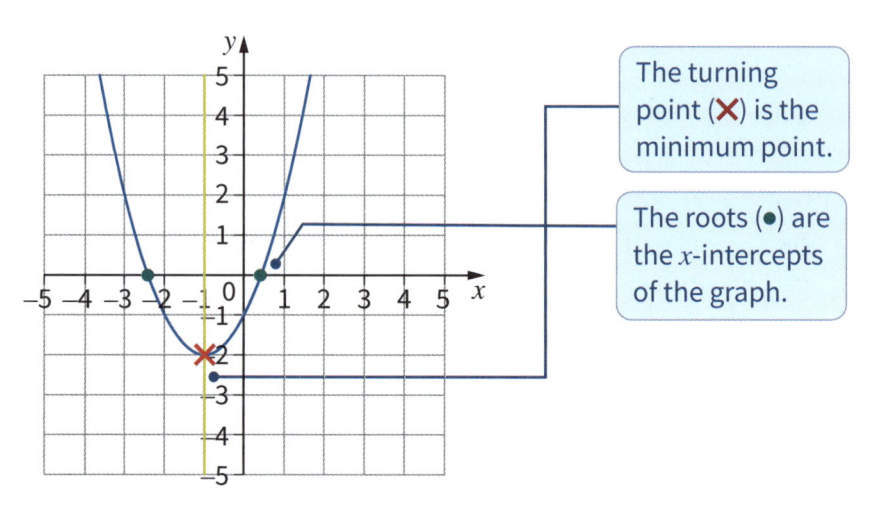

The turning point (✗) is the minimum point.

The roots (●) are the x-intercepts of the graph.

10

Quadratic graphs

Roots, intercepts, and turning points can also be found algebraically.

Worked example

$y = 2x^2 + 11x + 12$

(a) Find the coordinates of the points where the graph intersects the x- and y-axes.

$y = 2 \times 0^2 + 11 \times 0 = 12$,

y-intercept = **(0, 12)**

> The y-**intercept** occurs when $x = 0$. So substitute 0 for x and solve to find y.

$2x^2 + 11x + 12 = 0$
$2x^2 + 3x + 8x + 12 = 0$
$x(2x + 3) + 4(2x + 3) = 0$
$(x + 4)(2x + 3) = 0$

> The x-intercepts (roots) occur when $y = 0$. So set the equation equal to 0.

> Factorise the equation.

When $x + 4 = 0$, $x = -4$, so **(−4, 0)** is a root.
When $2x + 3 = 0$, $x = -1.5$, so **(−1.5, 0)** is a root.

> Set each factor equal to 0 and solve.

LINK
For a reminder of solving quadratic equations, look back at Chapter 9.

(b) Find the coordinates of the turning point of the graph.

$y = 2\left[x^2 + \frac{11}{2}x + 6\right]$

$= 2\left[\left(x + \frac{11}{4}\right)^2 - \left(\frac{11^2}{4^2}\right) + 6\right]$

$= 2\left[\left(x + \frac{11}{4}\right)^2 - \left(\frac{25}{16}\right)\right]$

$= 2\left(x + \frac{11}{4}\right)^2 - \left(\frac{25}{8}\right)$

$= 2\left(x + \frac{11}{4}\right)^2 > 0$ for all values of x so the minimum value of y is $-\frac{25}{8}$

> Find the lowest possible value for y by completing the square. Take out a factor of 2 (the coefficient of x^2).

> Halve the coefficient of x to become the constant term in the squared bracket.

> Combine the constant terms.

$\left(x + \frac{11}{4}\right)^2 = 0$

$x = -\frac{11}{4}$

> Now you have the minimum-possible y-value, find the corresponding x-value.

Turning point is at $\left(-\frac{11}{4}, -\frac{25}{8}\right) = (-2.75, -3.125)$

(c) Hence, **sketch** the graph of $y = 2x^2 + 11x + 12$

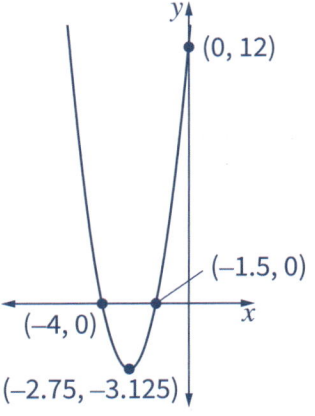

REVISION TIP

When you are asked to sketch a graph, you only need to show the shape, the coordinates where the line intersects the axes, and the turning point.

If you are asked to plot a graph, you need to use a table of values and plot it on graph paper.

10 Knowledge

Knowledge

10 Quadratic graphs, iterations, solving quadratic inequalities

Quadratic inequalities

Quadratic inequalities can also be solved algebraically.

Worked example

Solve $2x^2 - 5x \leq 12$, giving your answer in set notation.

$2x^2 - 5x - 12 \leq 0$ — Rearrange to get 0 on its own on one side of the equation.

$2x^2 - 5x - 12 = 0$

$(2x + 3)(x - 4) = 0$ — Rewrite as an equation and factorise.

$2x + 3 = 0$, so $2x = -3$
and $x = -\dfrac{3}{2}$
or
$x - 4 = 0$, so $x = 4$ — Set each factor equal to 0 and solve for x.

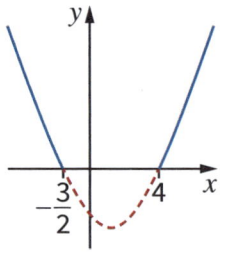

Draw a sketch to help. You are solving $2x^2 - 5x - 12 \leq 0$, so you need the part of the graph where $y \leq 0$ (shown in red).

$-\dfrac{3}{2} \leq x \leq 4$

in set notation: $\left\{x: -\dfrac{3}{2} \leq x \leq 4\right\}$ — Give your answer in set notation.

REVISION TIP

When sketching a graph to help solve a quadratic inequality, you only need to label the roots. These tell you the x-values for which the graph is positive and negative.

Worked example

$29 \leq \dfrac{p^2 + 9}{2} < 45$

Describe the range of values that p could take. Give your answer as an inequality.

LHS

$29 \leq \dfrac{p^2 + 9}{2}$

$58 \leq p^2 + 9$

$49 \leq p^2$

$7 \leq p, p \leq -7$

RHS

$\dfrac{p^2 + 9}{2} < 45$

$p^2 + 9 < 90$

$p^2 < 81$

$-9 < p < 9$

Solve each side as a separate inequality.

Sketching a graph might help.

$-9 < p \leq -7$

$7 \leq p < 9$ — Combine the two solutions.

72 10 Quadratic graphs, iterations, solving quadratic inequalities

10

Iterations

Iteration means repeating a process. It can be used to find an approximate solution to more complex equations.

To find the iterative formula for a cubic equation, e.g. $x^3 + bx^2 - c = 0$, rearrange the equation as follows:

$x_{n+1} = \sqrt[3]{c - (bx_n)^2}$

Substitute the initial value of x into the equation to find a solution. Then substitute the new value of x into the equation until you find two solutions that are the same to the specified accuracy (e.g. 2 d.p.). The solution lies between these two values.

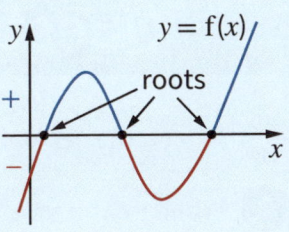

Worked example

(a) Show that the equation $x^3 + x^2 - 3 = 0$ has a solution between $x = 1$ and $x = 2$

$1^3 + 1^2 - 3 = -1$ — Substitute $x = 1$ into the equation.

$2^3 + 2^2 - 3 = 9$ — Substitute $x = 2$ into the equation.

The answers have different signs, so there must be a solution between $x = 1$ and $x = 2$ — Provide a rationale.

(b) Starting with $x_0 = 1$, use the iteration formula $x_{n+1} = \sqrt[3]{3 - (x_n)^2}$ to find an estimate, correct to 2 s.f. for a solution to $x^3 + x^2 - 3 = 0$

$x_1 = \sqrt[3]{3 - 1^2} = 1.2599\ldots$ — Substitute $x_0 = 1$ into the iteration formula.

$x_2 = \sqrt[3]{3 - 1.2599\ldots^2} = 1.1220\ldots$

$x_3 = \sqrt[3]{3 - 1.1220\ldots^2} = 1.2030\ldots$

Substitute your answer into the iteration formula, and repeat until you get two answers that round to the same number to 2 s.f.

$x_4 = \sqrt[3]{3 - 1.2030^2} = 1.1579\ldots$

x_4 and x_4 are both 1.2 to 2 s.f., so the estimate is $x = 1.2$ (2 s.f.).

> **REVISION TIP**
>
> You need to use your calculator efficiently for iteration questions. Start by typing in the value of x_0. For example, for part (b), first press [1]. Then press [=].
>
> Now enter the formula like this:
>
>   [3] [−] [ANS] [x^2]

Key terms — Make sure you can write a definition for these key terms.

iteration   line of symmetry   quadratic graph
roots   sketch   turning point

Retrieval

10 Quadratic graphs, iterations, solving quadratic inequalities

Learn the answers to the questions below, then cover the answers column with a piece of paper and write as many as you can. Check and repeat.

Questions | Answers

1. What is the general equation of a quadratic graph? — $y = ax^2 + bx + c$
2. What are the roots of a quadratic equation? — The x-intercept(s).
3. What is a turning point on a graph? — A maximum or minimum point.
4. Describe a quadratic graph with a negative x^2 term. — ∩-shaped curve.
5. When using an iterative formula, what does a change of sign between two x-values tell you? — That the equation has a root between those two x-values.
6. What should be labelled in a sketch of a quadratic graph? — The y-intercept, the x-intercept(s), and the turning point.
7. Where is the line of symmetry on a quadratic graph? — On the vertical line that passes through the turning point.
8. If the quadratic graph has one x-intercept, what does this mean? — There is one real root.
9. If a quadratic graph has no x-intercepts, what does this mean? — There are no real roots.
10. How do you solve a quadratic inequality? — Factorise or complete the square or use the quadratic formula. Sketch the curve. Determine the region of the graph that satisfies the inequalities.

Previous questions

Now go back and use these questions to check your knowledge of previous topics.

Questions | Answers

1. When you are asked to estimate the value of a calculation, what should you round each number in the calculation to? — 1 significant figure.
2. On a distance–time graph, what does a horizontal section represent? — No movement; the speed is 0.
3. How do you divide by a fraction? — Multiply by its reciprocal.
4. What is the general equation of a straight line? — $y = mx + c$
5. What is the general form of a quadratic expression? — $ax^2 + bx + c$

Practice

Exam-style questions

10.1 Solve each of these inequalities. Represent your solutions on the number lines.

(a) $x^2 - 4 \leq -3$

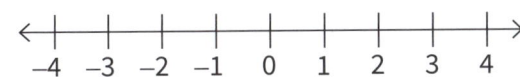

[3 marks]

(b) $7x^2 \geq 28$

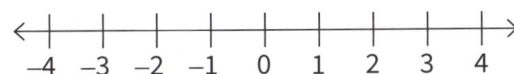

[3 marks]

10.2 Solve each of these inequalities. Write your answers using set notation.

(a) $x^2 - 8x + 15 \leq 0$ [4 marks]

(b) $3x^2 - x - 4 > 0$ [4 marks]

(c) $x^2 - 30 < -7x$ [4 marks]

10.3 n is an integer such that $2n + 7 \geq 3$ and $\dfrac{12-n}{n^2} > 1$

Find all possible values of n. [5 marks]

10.4 Solve $14 < \dfrac{w^2 - 7}{3} < 31$ [5 marks]

10.5 (a) Show that the equation $x^4 - 12x = 0$ has a solution between 2 and 3 [2 marks]

(b) Starting with $x_0 = 2$, use the iteration formula $x_{n+1} = \sqrt[4]{12x_n}$ to find an estimate for a solution of $x^4 - 12x = 0$ to 3 decimal places. [3 marks]

Exam-style questions

10.6 The graph with equation $y = x^3 + 5x^2 - 1$ is shown.

Dara wants to find the roots of the graph.

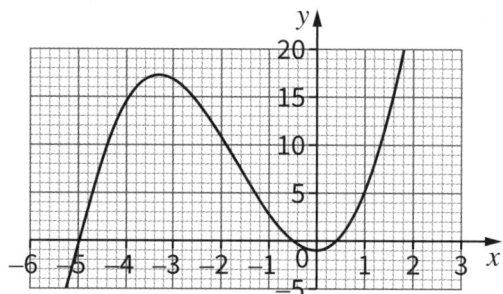

(a) Explain why Dara needs to solve the equation $x^3 + 5x^2 - 1 = 0$ **[1 mark]**

(b) Show that the equation $x^3 + 5x^2 - 1 = 0$ can be rearranged to
$x = \dfrac{1 - 5x^2}{x^2}$ **[2 marks]**

(c) Starting with $x_0 = -4$, use the iteration formula
$x_{n+1} = \dfrac{1 - 5(x_n)^2}{(x_n)^2}$
to find the value of a root of the graph.
Give your answer to 2 decimal places. **[3 marks]**

(d) Show that the equation $x^3 + 5x^2 - 1 = 0$ can be rearranged
to $x = \sqrt{\dfrac{1 - x^3}{5}}$ **[2 marks]**

(e) Starting with $x_0 = 0$, use the iteration formula
$x_{n+1} = \sqrt{\dfrac{1 - (x_n)^3}{5}}$
to find another root of the graph.
Give your answer to 2 decimal places. **[3 marks]**

(f) Starting with $x_0 = -1$, use the iteration formula
$x_{n+1} = \dfrac{1 - (x_n)^3}{5x_n}$
to find the final root of the graph.
Give your answer to 1 decimal place. **[3 marks]**

10.7 Given that $x^2 + 6x + 10 = (x + a)^2 + b$

(a) Find the value of a and the value of b. [2 marks]

(b) Hence write down the coordinates of the turning point on the graph $y = x^2 + 6x + 10$ [1 mark]

(c) Sketch the graph, clearly labelling the turning point and any intercepts. [2 marks]

Questions referring to previous content

10.8 Expand and simplify $(x - 1)(2 - x)(x + 4)$. [4 marks]

10.9 Solve the equation $x^2 - 25 = 6x$ by rearranging and completing the square. [3 marks]

Knowledge

11 Solving simultaneous equations

Simultaneous equations

In **simultaneous equations** the variables have the same values. You need to combine the equations to be able to find both values.

For example, for this pair of simultaneous equations:

$2x + y = 5$ and $x - y = 1$

the values $x = 2$ and $y = 1$ are true for both equations:

$2 \times 2 + 1 = 5$ and $2 - 1 = 1$

Solving simultaneous equations using graphs

Simultaneous equations can be solved using graphs.

The solution to a pair of simultaneous equations is the point at which their graphs **intersect**.

Worked example

Use a **graphical method** to find the solutions of the simultaneous equations:

$y = 2x + 7$ and $y = -x + 4$

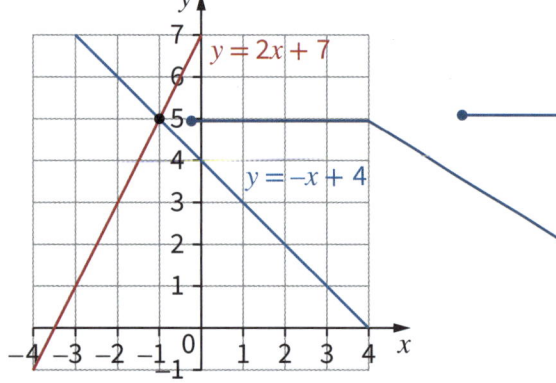

Draw both lines on the same axes.

$y = 2x + 7$ has gradient 2 and y-intercept 7

$y = -x + 4$ has gradient -1 and y-intercept 4

Find the point where the lines intersect. This is the solution to both equations.

The solution is $x = -1$, $y = 5$

> **REVISION TIP**
>
> You might need to rearrange the equations before you can plot them on the axes.

> **LINK**
>
> To remind yourself about drawing straight-line graphs, look back at Chapter 6.

Check that $x = -1$, $y = 5$ is a solution for both equations:

$y = 2x + 7$	$y = -x + 4$
$5 = (-1) + 7$	$5 = --1 + 4$
$5 = -2 + 7$ ✓	$5 = 1 + 4$ ✓

Solving simultaneous equations using graphs

Worked example

1. Use a graphical method to estimate, to 1 decimal place, the solutions to the simultaneous equations $y = x^2 - 2x$ and $y = x + 1$

x	−1	0	1	2	3
y	3	0	−1	0	3

Draw a table of values for the quadratic graph $y = x^2 - 2x$

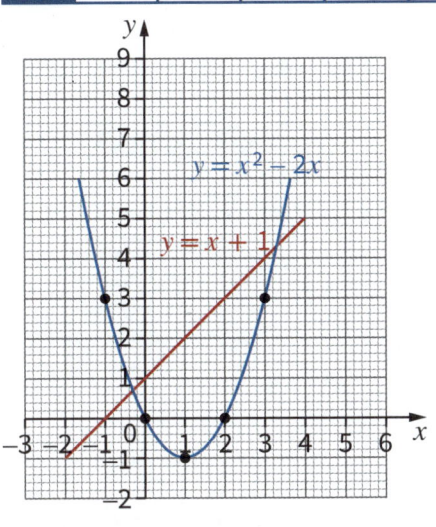

Plot the two graphs.

$y = x + 1$ is a straight line with gradient 1 and y-intercept 1.

The solutions are approximately
$x = -0.3$, $y = 0.7$ and $x = 3.3$, $y = 4.3$

Find the coordinates of the points where the two lines intersect. These are the solutions.

2. Using the graph of $y = x^2 - 2x$, estimate the solution to the equation $x^2 - 2x - 2 = 0$

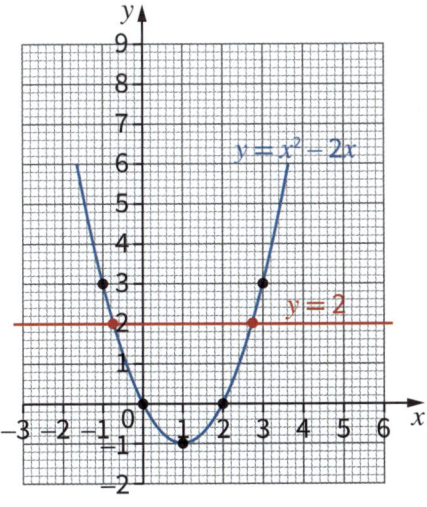

The graph of $y = x^2 - 2x$ has been given.

Add 2 to both sides to obtain the new equation:

$x^2 - 2x = 2$

Then draw on the graph the line $y = 2$.

Approximately $x = -0.7$ and $x = 2.7$

Use the intersection of the graphs to estimate the solutions.

Knowledge

11 Solving simultaneous equations

The substitution method

Sometimes, it is easier to **substitute** one variable into another; for example, if one equation has a variable as the subject.

Worked example

Find the values of x and y.

$y = x + 2$ ① $2x + 3y = 11$ ②

$2x + 3y = 11$ — Start with the more complicated equation.
$2x + 3(x + 2) = 11$ — Using ①, substitute $x + 2$ in place of y in equation ②.
$2x + 3x + 6 = 11$ — Multiply out the bracket.
$5x + 6 = 11$ — Collect like terms.
$5x = 5$ — Subtract 6 from both sides.
$x = 1$ — Divide by 5 to find the value of x
$y = x + 2$
$y = 1 + 2$ — Substitute $x = 1$ into equation ① to find y.
$y = 3$

REVISION TIP

You can check your answers by substituting $x = 1$, $y = 3$ into equation ②:
$2x + 3y = 2 \times 1 + 3 \times 3 = 2 + 9 = 11$ which is correct.

The elimination method

There are two **algebraic methods** to solve simultaneous equations. This means solving them using algebra and not a graph.

In the **elimination method** you add or subtract the two equations to get rid of either x or y.

Worked example

Solve this pair of simultaneous equations: $2x + 3y = 25$ and $6x - 2y = -2$

$2x + 3y = 25$ ①
$6x - 2y = -2$ ② — Number the equations ① and ②.

$(2x + 3y = 25) \times 3 = 6x + 9y = 75$ ③ — We need the x or y coefficient to be the same. Here, $6x$ is a multiple of $2x$, so multiply equation ① by 3. Label this equation ③.

$6x + 9y = 75$ ③
$-(6x - 2y = -2)$ ②
$\overline{0x + 11y = 77}$
$y = 7$ — Both equations now have $6x$, so we can eliminate x by subtracting equation ② from ③. Be careful subtracting negatives.

$2x + 3(7) = 25$
$2x + 21 = 25$ — Substitute $y = 7$ into equation ① to find x.
$2x = 4$
$x = 2$

The elimination method

Worked example

Check:

$2(2) + 3(7) = 25$ $6(2) - 2(7) = -2$

$4 + 21 = 25$ $12 - 14 = -2$

$25 = 25$ ✓ $-2 = -2$ ✓

REVISION TIP

If the signs are the same: **Subtract**

$5x \; \boxed{+ 2y} = 16$ ①
$-(3x \; \boxed{+ 2y} = 12)$ ②
$\overline{2x + 0y = 4}$

If the signs are different: **Add**

$5x \; \boxed{+ 2y} = 16$ ①
$+(3x \; \boxed{- 2y} = 12)$ ②
$\overline{8x + 0y = 28}$

Word-based problems

Sometimes you will be given a **word-based problem** from which you need to form two simultaneous equations.

Worked example

The cost of three adult tickets and two child tickets to a theme park is £155.

The cost of two adult tickets and five child tickets is £195.

1. Form a pair of simultaneous equations to describe this situation.

   > Decide what the two variables represent and give them each a letter.

   a = cost of adult ticket
   c = cost of child ticket.

   > Use the information in the question to form two equations using your variables. Number them ① and ② to help with the next part of the question.

   $3a + 2c = 155$ ① and $2a + 5c = 195$ ②

2. Solve your equations to find the cost of an adult ticket and the cost of a child ticket.

   > Multiply ① by 2: $6a + 4c = 310$ ③
   > Multiply ② by 3: $6a + 15c = 585$ ④
   > Both of the $6a$ terms are positive, so subtract equation 3 from equation 4.

   ④ − ③: $11c = 275$
   $c = 25$

   > Substitute into equation ①: $3a + 50 = 155$

   $3a = 105$, so $a = 35$

   The cost of an adult ticket is £35 and the cost of a child ticket is £25.

Key terms Make sure you can write a definition for these key terms.

algebraic method   elimination method   graphical method
intersect   simultaneous equations   substitute
word-based problem

Retrieval

11 Solving simultaneous equations

Learn the answers to the questions below, then cover the answers column with a piece of paper and write as many as you can. Check and repeat.

Questions | Answers

#	Question	Answer
1	What are simultaneous equations?	A pair of equations with two variables that you solve at the same time.
2	What are the two algebraic methods you can use to solve simultaneous equations?	Elimination and substitution.
3	What does the point of intersection represent on two graphs?	The solution to the pair of simultaneous equations.
4	To form simultaneous equations from word-based problems, what is the first step?	Decide what the two variables represent.
5	How can you check your solutions to a pair of simultaneous equations?	Substitute your values back into the equations.
6	How would you eliminate x from the equations $3x + 5y = 1$ ① $3x - 2y = 8$ ②	Subtract equation ② from equation ①.
7	How would you eliminate y from the equations $2x - y = 2$ ① $7x + y = 43$ ②	Add the two equations together.
8	What would be your first step if you were asked to solve these simultaneous equations by elimination? $2x + 3y = 7$ ① $x - 2y = 4$ ②	Multiply one or both equations so that either x or y have the same coefficient in both equations.
9	If you form and solve simultaneous equations from a word problem, what should your final step be?	Use your solutions to answer the original question in context.

Previous questions

Now go back and use these questions to check your knowledge of previous topics.

Questions | Answers

#	Question	Answer
1	What does $a \geq b$ mean?	a is greater than or equal to b.
2	What is an identity?	An equation that is true for any value of the variable (s).
3	When a number is written in standard form $(a \times 10^n)$ what values can the index number, n, take?	Any integer.
4	What does it mean to expand a single bracket?	To multiply each term inside the bracket by the term in front of the bracket.

Practice

Exam-style questions

11.1 The graphs of $y = 2x - 3$ and $x + y = 4$ are shown.

Use the graphs to estimate the solution to these simultaneous equations.

$y = 2x - 3$

$x + y = 4$

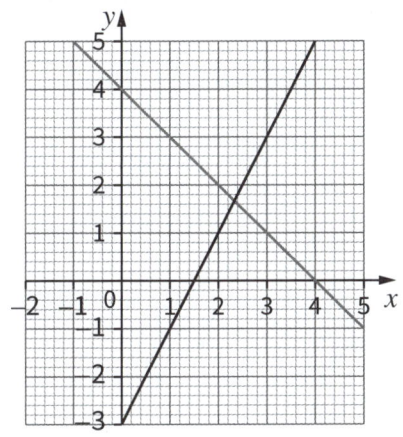

EXAM TIP
The solution of two simultaneous equations is where the lines intersect.

[2 marks]

11.2 The graph of $x + 2y = 6$ is shown.

By drawing a suitable line on the graph, find the solution to these simultaneous equations.

$x + 2y = 6$

$y = x$

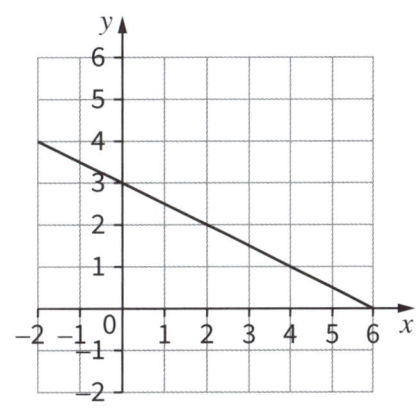

[2 marks]

11.3 Solve each pair of simultaneous equations.

(a) $x + y = 14$ $\qquad$ $x - y = 8$ $\qquad$ [3 marks]

(b) $2x - 2y = 4$ $\qquad$ $2x + 3y = 14$ $\qquad$ [3 marks]

(c) $4x + 5y = 37$ $\qquad$ $2x + y = 11$ $\qquad$ [3 marks]

(d) $3x - 2y = 2$ $\qquad$ $12x - 4y = 10$ $\qquad$ [3 marks]

11.4 Solve each pair of simultaneous equations.

(a) $2x + 5y = 11$ $\qquad$ $3x - 2y = -12$ $\qquad$ [3 marks]

(b) $2x - 7y = 12$ $\qquad$ $5x - y = -3$ $\qquad$ [3 marks]

(c) $3x + 8y = 12$ $\qquad$ $2x + 12y = 13$ $\qquad$ [3 marks]

(d) $6x - 4y = 9$ $\qquad$ $5x + 3y = -2$ $\qquad$ [3 marks]

EXAM TIP
You might need to multiply both equations before you can eliminate x or y.

Exam-style questions

11.5 Two families go to the cinema.

The first family buys one adult ticket and three child tickets and pays £39

The second family buys two adult tickets and four child tickets and pays £62

    (a) Form a pair of simultaneous equations to describe this situation. **[2 marks]**

    (b) Solve your equations to find the cost of an adult ticket and the cost of a child ticket. **[3 marks]**

11.6 The mass of 20 apples and 30 satsumas is 4050 g.

The mass of 12 apples and 15 satsumas is 2205 g.

Find the mass of one apple and the mass of one satsuma. **[4 marks]**

> **EXAM TIP**
> Start by writing a pair of simultaneous equations to represent the information.

11.7 $2^x \times 2^y = 64$ and $2^x \div 2^y = 4$

Find the values of x and y. **[4 marks]**

> **EXAM TIP**
> Use the laws of indices to write two linear simultaneous equations.

11.8 Use a graphical method to find the solutions to the simultaneous equations

$y = (x - 2)^2$

$y = x$

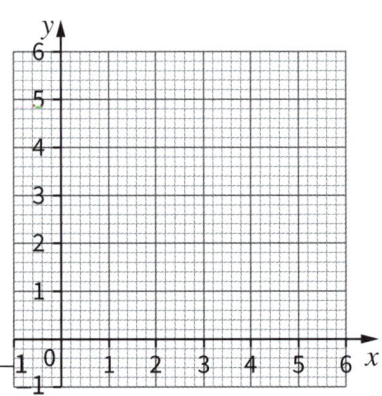

[4 marks]

11.9 Use an algebraic method to solve these pairs of simultaneous equations.

    (a) $x + y = 3$ $y = x^2 + 3x - 2$ **[4 marks]**

    (b) $2x^2 + y^2 = 9$ $2y + x = 0$ **[4 marks]**

> **EXAM TIP**
> Be careful when squaring. Look at the coefficients too.

11.10 The graph of $y = x^2 - 3x - 18$ is shown.

(a) Use a graphical method to find the solutions to the simultaneous equations

$y = x^2 - 3x - 18$

$y = 2x - 12$ [3 marks]

(b) Use the graph to estimate the solutions to the equation
$x^2 - 3x - 14 = 0$ [3 marks]

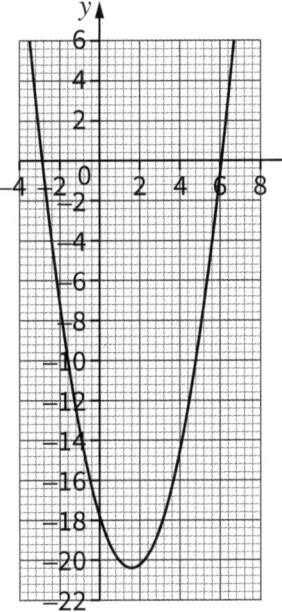

Questions referring to previous content

11.11 Solve the inequality for x and show your solution on the number line.

$11 + 2x \geq 5x - 1$

[2 marks]

11.12 Solve $6x^2 + 7x - 20 > 0$, giving your answer in set notation. [4 marks]

Knowledge

12 Sequences

Sequences: an overview

A **sequence** is an ordered set of numbers or patterns that follow the same rule.

The numbers in a sequence are called **terms**.
first number = first term
second number = second term
nth number = nth term

An **arithmetic sequence** (or linear sequence) is one in which the next term can be found by adding or subtracting the same number each time.

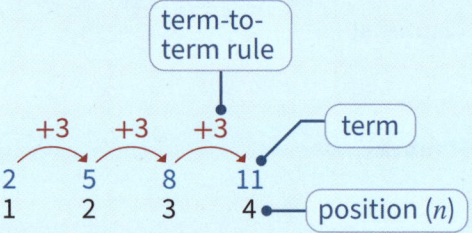

Term-to-term rule

The rule that gets you from one term to the next. This is the same for every term.

Position-to-term rule

A rule that finds the term based on where it is in the sequence. The position to term rule is always written in terms of n. This is called the nth term. In the example above, it is $3n - 1$.

The term-to-term rule

Worked example

The first three patterns in a sequence made from sticks are shown.

Pattern 1   Pattern 2   Pattern 3

1. Write down the term-to-term rule for this sequence.

   Make a table and count the number of sticks for each pattern.

n	1	2	3
sticks	3	5	7

   (+2, +2)

   Two more sticks are added each time. So this is an arithmetic sequence.

   The term-to-term rule is 'add 2'.

2. How many sticks will there be in Pattern 6?

   Continue adding 2.

n	1	2	3	4	5	6
sticks	3	5	7	9	11	13

   (+2, +2, +2, +2, +2)

   There are 13 sticks in Pattern 6.

The position-to-term rule

Worked example

The nth term of a sequence is $11 + 3n$.

1. Work out the seventh term of the sequence.

   seventh term
   $= 11 + 3 \times 7 = 32$ ← *Substitute $n = 7$*

   REVISION TIP

   The question gives you an expression for the nth term, so you know that $11 + 3n$ is the position-to-term rule.

2. Is 81 a term of this sequence? Show how you get your answer.

   $11 + 3n = 81$
   $$\frac{3n}{3} = \frac{70}{3}$$
   $n = 23.\dot{3}$

   70 is not a multiple of 3, so the solution is not an integer. Hence 81 is **not** in the sequence.

12

Finding the nth term

The **nth term** is the position-to-term rule. You can use it to find any term in the sequence. n is the term number. To find the nth term, follow the steps in the worked example.

Worked example

Find the nth term of the sequence: 5, 9, 13, 17, ...

5 9 13 17
 +4 +4 +4

← Work out the common difference between terms.

Common difference = 4

$4n$: 4 8 12 16

← Write out the multiples of the common difference.

↓+1 ↓+1 ↓+1 ↓+1
 5   9   13 17

← Work out what to add or subtract to get the original sequence.

nth term = $4n + 1$

$(4 \times 2) + 1 = 9$

← Write out the nth term rule.

← Check your answer by substituting in $n = 2$.

Special sequences

Special sequences do not have a common difference between their terms. They are not arithmetic sequences.

Fibonacci sequences

Any sequence where you add the previous two terms to get the next term is a Fibonacci-*type* sequence. The 'original' **Fibonacci sequence** was:

1, 1, 2, 3, 5, 8, 13, 21 ...

Worked example

The first three terms of a Fibonacci-type sequence are $k, k + 6, 2k + 6$

Work out the fourth term of the sequence.

$= (k + 6) + (2k + 6)$
$= 3k + 12$

← 4th term = 2nd term + 3rd term

The first five **square numbers**:

 $2^2 = 4$

 $3^2 = 9$

 $4^2 = 16$

 $5^2 = 25$

 $6^2 = 36$

The first three **cube numbers**:

 $1^3 = 1$

 $2^3 = 8$

 $3^3 = 27$

The first four **triangular numbers**:

1

$1 + 2 = 3$

$1 + 2 + 3 = 6$

$1 + 2 + 3 + 4 = 10$

REVISION TIP

To find the next triangular number, just add on the next positive integer.

12 Knowledge 87

Knowledge

12 Sequences

Quadratic sequences

In a **quadratic sequence**, the position-to-term rule includes a squared term.

> **Worked example**
>
> 1. Here are the first five terms of a sequence.
>
>   1, 5, 12, 22, 35
>
>   Find the next term of this sequence.
>
>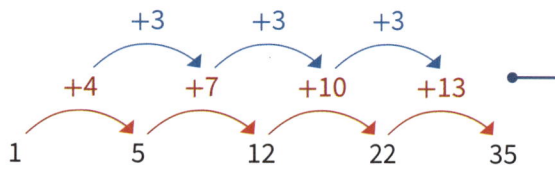
>
>   Calculate:
>   - the difference between each term
>   - the difference between the first differences.
>
>   next difference: $13 + 3 = 16$
>   next term: $35 + 16 = 51$
>
>   Find the difference to the next term. Use this to find the next term.
>
> 2. The nth term of a different sequence is $n^2 - 2n$.
>
>   What is the 3rd term of the sequence?
>
> $n^2 - 2n$
>
>   3rd term $= 3^2 - (2 \times 3)$
>
> $= 3$
>
>   To find the 3rd term of the sequence, substitute $n = 3$

> **Worked example**
>
> Part of a quadratic sequence is: 4, 5, 12, 25, …
>
> Work out an expression for the nth term.
>
>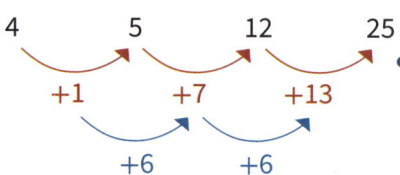
>
> Find the second difference.
>
> Sequence involves $3n^2$.
>
> Halve the second difference and multiply it by n^2.
>
> | n | 1 | 2 | 3 | 4 |
> |---|---|---|---|---|
> | | 4 | 5 | 12 | 25 |
> | $3n^2$ | 3 | 12 | 27 | 48 |
> | | 1 | −7 | −15 | −23 |
>
> Subtract $3n^2$ from each term of the original sequence.
>
> Find the nth term of the new linear sequence: 1, −7, −15, −23
>
> | n | 1 | 2 | 3 | 4 |
> |---|---|---|---|---|
> | | 1 | −7 | −15 | −23 |
>
> difference between terms is −8
>
> nth term is $-8n + 9$
>
> Put this together with the quadratic term you found earlier.
>
> The nth term of the original sequence is $3n^2 - 8n + 9$
>
> **REVISION TIP**
>
> Check the nth term by substituting in a value for n.

12

Geometric sequences

The next term in a **geometric sequence** is found by multiplying or dividing by the same number each time. For example, the term-to-term rule for this sequence is '×3'

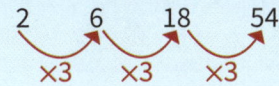

Worked example

1. A sequence has nth term rule 8^n

(a) Work out the third term of the sequence.

$8^3 = 512$ ← Substitute 3 in place of n

(b) What is the term-to-term rule?

Multiply by 8

To find the term-to-term rule, divide a term by the preceding term. Check it holds true for at least 2 instances.

2. A sequence has nth term rule $(\sqrt{5})^n$

(a) Work out the fourth term of the sequence.

$(\sqrt{5})^4$
$= (5^{\frac{1}{2}})^4$
$= 5^2$
$= 25$

Substitute 4 in place of n, then simplify the surd.

(b) What is the term-to-term rule?

The term-to-term rule is 'multiply by $\sqrt{5}$'

LINK
To remind yourself about surds, look back at Chapter 3.

Worked example

The first three terms of a geometric sequence are $(\sqrt{x} - 2)$, 1, $(\sqrt{x} + 2)$

(a) Find the term-to-term rule.

$$\frac{1}{(\sqrt{x} - 2)} = \frac{(\sqrt{x} + 2)}{1}$$

Use the fact that the second term divided by the first term is equal to the third term divided by the second term to write an algebraic equation.

$(\sqrt{x} - 2)(\sqrt{x} + 2) = 1$
$x - 2\sqrt{x} + 2\sqrt{x} - 4 = 1$
$x - 4 = 1$
$x = 5$

Multiply both sides by $(\sqrt{x} - 2)$ then, expand the brackets and solve it to find the value of x.

Find the term-to-term rule by dividing the third term by the second term and substitute the value for x.

The term-to-term rule is 'multiply by $(\sqrt{5} + 2)$'.

(b) Show that the 4th term is $4\sqrt{5} + 9$

4th term $(\sqrt{5} + 2)(\sqrt{5} + 2)$
$= 5 + 2\sqrt{5} + 2\sqrt{5} + 4$
$= 4\sqrt{5} + 9$

Substitute '$x = 5$' into third term then multiply it by the term-to-term rule found in part (a).

Expand and simplify.

Key terms — Make sure you can write a definition for these key terms.

arithmetic sequence   cube numbers   geometric sequence
Fibonacci sequence nth term   position-to-term rule
quadratic sequence   term   term-to-term rule
triangular numbers   sequence   square numbers

Retrieval

12 Sequences

Learn the answers to the questions below, then cover the answers column with a piece of paper and write as many as you can. Check and repeat.

	Questions	Answers
1	What is an arithmetic sequence?	A sequence that increases or decreases by the same amount each time.
2	How do you get to the next term of a geometric sequence?	Multiply or divide by the same number each time.
3	What is the term-to-term rule?	The rule that gets you from one term to the next. This is the same for every term.
4	What can you use the nth term of a sequence for?	To find any term in the sequence.
5	Is 14 in the sequence $3n - 1$?	Let $14 = 3n - 1$ So $15 = 3n \Rightarrow n = 5$ Since $n = 5$ is an integer, 14 is in the sequence.
6	How do you find the next term in a Fibonacci sequence?	Add the previous two terms to get the next one.
7	Is 15 a triangular number? How do you know?	Yes, because $1 + 2 + 3 + 4 + 5 = 15$.
8	What is a cube number?	A number multiplied by itself three times, such as $3 \times 3 \times 3 = 27$.
8	What is the general form of a quadratic sequence?	$ax^2 + bx + c$
9	How do you find the common ratio of a geometric sequence?	Divide any term by the previous term.
10	What is a linear sequence?	A sequence that increases or decreases by the same amount.
11	What are the first five terms of the square number sequence?	1, 4, 9, 16, 25
12	True or false? Sequences always involve numbers.	False; you could be asked to draw a sequence using patterns.

Previous questions

Now go back and use these questions to check your knowledge of previous topics.

	Questions	Answers
1	How do you represent an inequality on a number line where the value is not included?	With a hollow circle.
2	How do you find 50% of a number?	Divide by 2 or multiply by 0.5.
3	How do you write a ratio in its simplest form?	Divide by the highest common factor.

Exam-style questions

12.1 Complete the table for each sequence. The first row has been done for you.

	Position-to-term rule (nth term)	First four terms	Term-to-term rule	7th term	100th term	Type of sequence
e.g.	$2n - 1$	1, 3, 5, 7	Add 2	13	199	Arithmetic
(a)	$5n + 2$					
(b)	5×2^n					
(c)	$n^2 - 2n$					
(d)		1, 1, 2, 3				

[18 marks]

12.2 The nth term of a sequence is $8n + 3$.

(a) Which term in the sequence is 51? [2 marks]

(b) Show that 64 is not a number in the sequence. [2 marks]

(c) Find the first number in the sequence to exceed 100. [3 marks]

EXAM TIP
Write and solve an inequality.

12.3 The nth term of a sequence is $n^2 - 30$.

(a) Find the 4th number in the sequence. [2 marks]

(b) Is the number 114 in this sequence?

Explain your answer. [2 marks]

12.4 A Fibonacci sequence starts $m, n, m + n, ...$

(a) Write an algebraic expression for

  (i) the 4th term in the sequence [1 mark]

  (ii) the 7th term in the sequence. [2 marks]

(b) The 1st term of the sequence is 3.

The difference between the 1st and 3rd terms is 5.

Work out the 8th term of the sequence. [2 marks]

12.5 Complete the table for each sequence. The first one has been done for you.

	Sequence	Term-to-term rule	Position-to-term rule (nth term)	10th term
e.g.	2, 6, 10, 14	Add 4	$4n - 2$	$(4 \times 10) - 2 = 38$
(a)	17, 23, 29, 35			
(b)	−1, 2, 5, 8			
(c)	4, 1, −2, −5			
(d)	20, 15, 10, 5			
(e)	3, 3.5, 4, 4.5			

[15 marks]

Exam-style questions

12.6 An arithmetic sequence has first term 5 and third term 11.

Find an expression, in terms of n, for the nth term of the sequence. **[3 marks]**

EXAM TIP
Work out some terms in the sequence first.

12.7 Calculate the sum of the 50th and 60th terms in the sequence that starts

12, 9, 6, 3, … **[3 marks]**

12.8 Here is a number sequence: $3, \dfrac{5}{2}, \dfrac{7}{3}, \dfrac{9}{4}, \dfrac{11}{5}, \ldots$

(a) Find the next term in the sequence. **[1 mark]**

(b) Find an expression, in terms of n, for the nth term of the sequence. **[3 marks]**

(c) Find the product of the 6th and 9th terms of the sequence. **[2 marks]**

EXAM TIP
Look at the numerators and denominators separately.

12.9 Write down the first five terms for each quadratic sequence.

(a) $n^2 + 4$ **[2 marks]**

(b) $2n^2 + 5n - 6$ **[2 marks]**

(c) $17 + 32n - 5n^2$ **[2 marks]**

12.10 Find an expression, in terms of n, for the nth term of these quadratic sequences.

(a) 1, 4, 9, 16, … **[1 mark]**

(b) 0, 3, 8, 15, … **[1 mark]**

(c) 3, 12, 27, 48, … **[1 mark]**

12.11 Find the next two terms in these sequences.

(a) 5, 8, 13, 20, … **[2 marks]**

(b) −2, 5, 14, 25, … **[2 marks]**

12.12 A sequence has nth term $n^2 + 2n + 2$

Work out which term in the sequence has a value of 50. **[3 marks]**

EXAM TIP
Write and solve an equation.

12.13 Find the nth term of these sequences.

(a) 10, 12, 16, 22, … **[3 marks]**

(b) −9, 2, 17, 36, … **[3 marks]**

12.14 Rachel thinks that every term in the sequence $n^2 + 4n + 6$ is positive.

Explain how you know Rachel is correct. [1 mark]

12.15 The nth term of sequence is given by $an^2 + b$, where a and b are integers.

The 4th term of the sequence is 42.

The 9th term of the sequence is 237.

Find the 15th term in the sequence. [5 marks]

12.16 The first three terms of a geometric sequence are

$9 + \sqrt{y}$, 4, $9 - \sqrt{y}$

Find the value of y.

Show your working. [3 marks]

12.17 (a) A sequence has the nth term $\dfrac{n}{n+3}$

Write the first three terms of the sequence. [2 marks]

(b) Hence, write an expression, in terms of n, for the nth term of the sequence

$\dfrac{3}{5}, \dfrac{4}{7}, \dfrac{5}{9}, \dfrac{6}{11}, \dfrac{7}{13}, \dots$

[2 marks]

12.18 (a) A sequence has the nth term $\dfrac{(\sqrt{3})^n}{3}$

Write the first three terms of the sequence. [3 marks]

(b) Write an expression, in terms of n, for the nth term of the sequence

$5\sqrt{2}, 10, 10\sqrt{2}, 20, 20\sqrt{2}, \dots$ [2 marks]

Questions referring to previous content

12.19 Solve this pair of simultaneous equations: [4 marks]

$3y - 4x = 18$ and $10x = y - 32$

12.20 Harriet buys three pineapples and six bananas for £17.10

Eric buys four pineapples and 9 bananas for £24.05

Work out the cost of one pineapple and one banana. [5 marks]

Knowledge

13 Cubic graphs, reciprocal graphs, exponential graphs, transformation of graphs

Cubic and reciprocal graphs

A **cubic graph** has an x^3 term in its equation.

For example, $y = x^3 + x^2 + 4x + 7$

The general equation of a cubic graph is:

$y = ax^3 + bx^2 + cx + d$

where a, b, c, and d are constants.

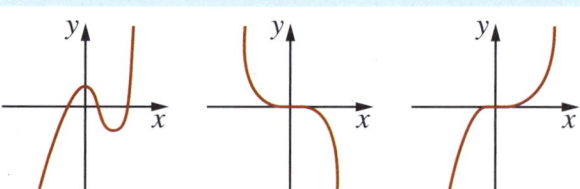

A **reciprocal graph** is of the form $y = \dfrac{a}{x}$, where a is a constant.

When $y = \dfrac{1}{x}$, the value of x cannot be 0, because $\dfrac{1}{0}$ is not a defined value.

The curve will get closer and closer to the axes but will never touch them.

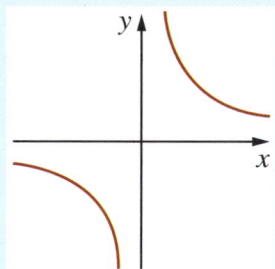

REVISION TIP

An inverse proportion graph is a type of reciprocal graph. You will revise inverse proportion in Chapter 17.

Worked example

Match each graph with a possible equation.

$y = 2x - 1$ $y = 2x^2 - 1$ $y = 2x^3 - 1$ $y = \dfrac{2}{x}$

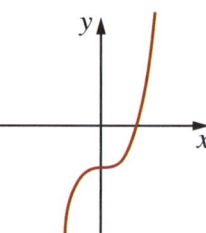

Graph is a cubic
$y = 2x^3 - 1$

(quadratic graph)

Graph is a quadratic
$y = 2x^2 - 1$

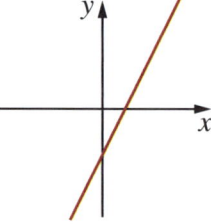

Graph is a straight line
$y = 2x - 1$

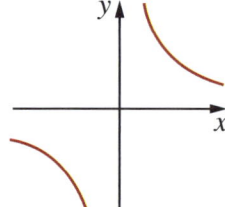

Graph is a reciprocal
$y = \dfrac{2}{x}$

Worked example

1. Complete the table of values for $y = x^3 + 1$

x	-2	-1	0	1	2
y	-7	0	1	2	9

2. Draw the graph of $y = x^3 + 1$

It is cubic, so we expect a ∿ shape.

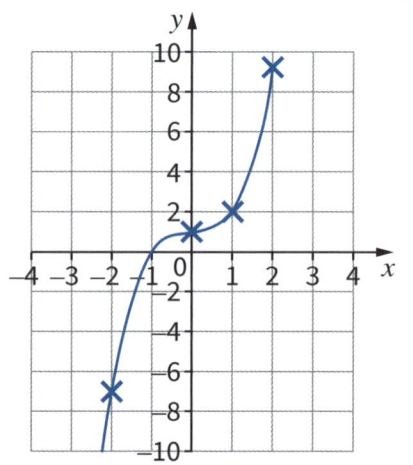

REVISION TIP

If it's a cubic graph, look for x^3 in the equation.

If it's a quadratic graph, look for x^2 in the equation.

13

Cubic and reciprocal graphs

> **Worked example**
>
> The time, t hours, taken to complete a journey at speed v km/h is given by the equation $t = \dfrac{75}{v}$.
>
> 1. Find the missing values for t and v.
>
> | v (km/h) | 1 | 3 | 5 | $\dfrac{75}{7.5} = 10$ |
> |---|---|---|---|---|
> | t (hours) | $\dfrac{75}{1} = 75$ | 25 | $\dfrac{75}{5} = 15$ | 7.5 |
>
> 2. Plot a graph of t against v.
>
>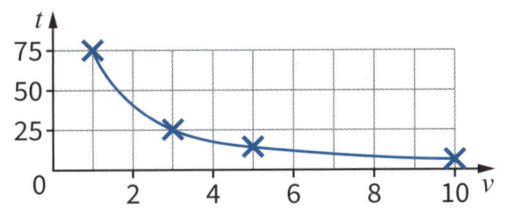

> **REVISION TIP**
>
> Start by drawing a sketch of the type of graph you expect, so that you can check it once you have plotted it.

Exponential graphs

Functions where the variable is in the power are called **exponential** functions.

An exponential function is in the form $y = k^x$, where k is a positive constant.

The graph of $y = k^x$ has y-intercept 1 (since $k^0 = 1$ for any value of k).

$k > 1$ indicates exponential growth

$k < 1$ indicates exponential decay

The graph of $y = Ak^x$ will have y-intercept A.

> **Worked example**
>
> The number, N, of bacteria in a Petri dish at the start of an experiment is 300. The number of bacteria in the dish after t hours is given by $N = A \times 1.05^t$, where A is a constant.
>
> **(a)** Work out the value of A.
>
> $A \times 1.05^0 = 300$ — Initially, $t = 0$ and $N = 300$
>
> $A = 300$ since $1.05^0 = 1$ — Substitute into the equation.
>
> **(b)** How many bacteria will be in the dish after seven hours?
>
> number of bacteria $= 300 \times 1.05^7 = 422$ — Substitute into the equation.
>
> **(c)** Sketch the graph of N against t.
>
> Sketch an exponential graph, with y-intercept of 300. Make sure you label the axes.

Knowledge

13 Cubic graphs, reciprocal graphs, exponential graphs, transformation of graphs

Graph transformations

The graph of $y = f(x)$, can be **translated** in the following ways:

- $y = f(x + a)$ is a translation of a units left
- $y = f(x - a)$ is a translation of a units right
- $y = f(x) + a$ is a translation of a units up
- $y = f(x) - a$ is a translation of a units down

REVISION TIP

For $y = f(x)$
- the x is inside the brackets, so $f(x + a)$ will change the x-coordinate
- the y is outside the brackets, so $f(x) + a$ will change the y-coordinate.

WATCH OUT

Pay attention to the positive and negative signs, and the impact they have on whether the graph moves to the left, right, up or down.

LINK

You will revise **transformations** and **translations** in Chapter 23.

Worked example

The graph of $y = f(x)$ is shown.

Sketch the graph of $y = f(x - 2)$.

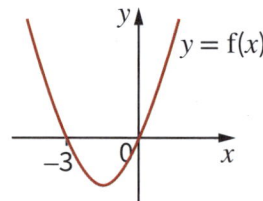
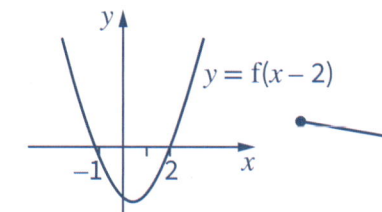

The graph is translated 2 units to the right. Add 2 to each of the x-intercepts.

Worked example

The graph of $y = f(x)$ is shown.

Sketch the graph of $y = f(x) + 2$.

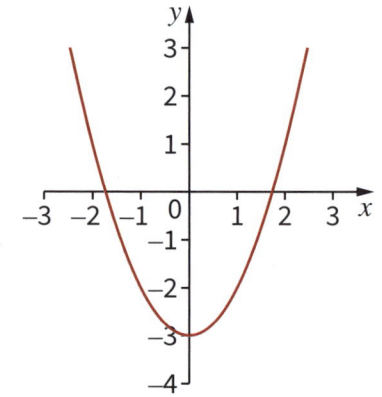
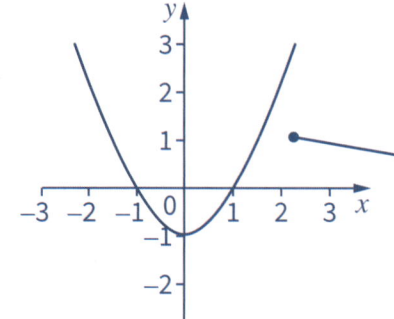

The graph is translated 2 units up. Add 2 onto the y-intercept.

Graph transformations

If you have the graph of $y = f(x)$, then:
- $y = -f(x)$ is a reflection in the x-axis
- $y = f(-x)$ is a reflection in the y-axis

REVISION TIP

Remember this...

The x is inside the brackets, so $f(-x)$ will change the x-coordinate.

The y is outside the brackets, so $-f(x)$ will change the y-coordinate.

Worked example

The graph of $y = f(x)$ is shown.

Sketch the graph of $y = -f(x)$.

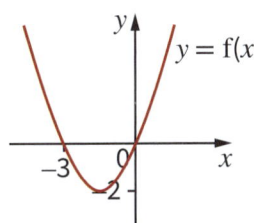
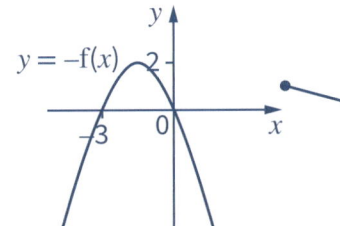

The graph is reflected in the x-axis.
The x-intercepts stay the same.
Only the y-coordinates are affected.

Worked example

The graph of $y = f(x)$ is shown.

Sketch the graph of $y = f(-x)$.

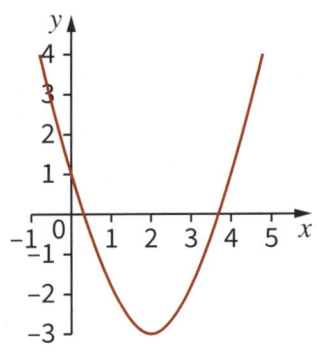
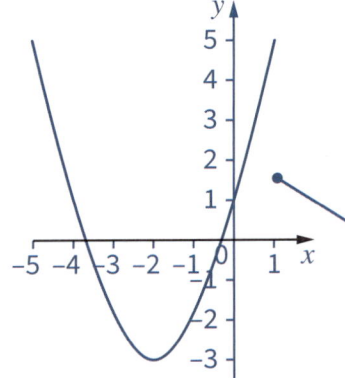

The graph is reflected in the y-axis.
The y-intercepts stay the same.
Only the x-coordinates are affected.

Worked example

The graph of the curve C, with equation $y = f(x)$, is transformed to give the graph of the curve D, with equation $y = -f(x + 5)$.

The point on C with coordinates $(3, 5)$ is mapped to the point R on curve D.

Find the coordinates of point R.

x-coordinate: $3 - 5 = -2$

y-coordinate: $5(-1) = -5$

The coordinates of R are $(-2, -5)$

The 5 inside the brackets tells you to subtract 5 from the x-coordinate.

The negative sign outside the brackets tells you to multiply the y-coordinate by -1

Key terms — Make sure you can write a definition for these key terms.

cubic graph   exponential   reciprocal graph
transformation   translation

Retrieval

13 Cubic graphs, reciprocal graphs, exponential graphs, transformation of graphs

Learn the answers to the questions below, then cover the answers column with a piece of paper and write as many as you can. Check and repeat.

#	Questions	Answers
1	What is the general equation of a cubic graph?	$y = ax^3 + bx^2 + cx + d$
2	What are the roots of a cubic equation?	The x-intercepts.
3	What is the general equation for a reciprocal graph?	$\dfrac{a}{x}$
4	How does the equation differ between exponential decay and growth?	Exponential decay, k < 1 and exponential growth, k > 1
5	What is an exponential function?	A function of the form AK^x.
6	What is the y-intercept of $y = k^x$?	$y = 1$, as $k^0 = 1$
7	What transformation of f(x) does $y = $ f($x + a$) represent?	Translation of a units left.
8	What transformation of f(x) does $y = $ f(x) $+ a$ represent?	Translation of a units up.
9	What transformation of f(x) does $y = -$f(x) represent?	Reflection in the x-axis.
10	What transformation of f(x) does $y = $ f($-x$) represent?	Reflection in the y-axis.

Previous questions

Now go back and use these questions to check your knowledge of previous topics.

#	Questions	Answers
1	When you see a real-life graph, what should you do first?	Find out what each unit on each scale represents.
2	If there are 10 squares between 0 and 20, what does each square represent?	2 units.
3	On a distance–time graph, what does the gradient represent?	Speed.
4	On a velocity–time graph, what does the gradient represent?	Acceleration.
5	On a distance–time graph, what does a horizontal section represent?	No movement.

Practice 13

Exam-style questions

13.1 (a) Nazia says that the inverse of a function f(x) is $\frac{1}{f(x)}$.

Explain Nazia's misunderstanding. **[1 mark]**

(b) Find the inverse function of f(x), where
$f(x) = \frac{1-x}{2x+4}, x \neq -2$ **[3 marks]**

13.2 The sketch shows the graph of $y = ab^x$, where $a > 0$ and $b > 0$.
The graph passes through the points (1, 10) and (−1, 0.4).
Find the values of a and b.

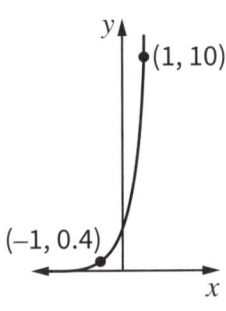

[3 marks]

13.3 Here are four graphs.

Graph A   Graph B   Graph C   Graph D

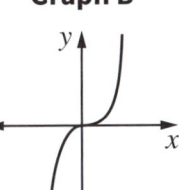 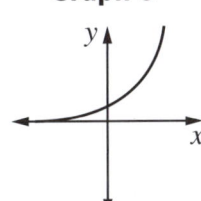 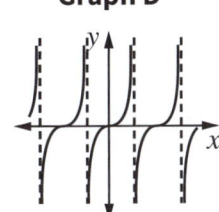

The graphs represent four different types of function.

Match each description of the function in the table to the letter of the graph.

Description of function	Graph
Exponential	
Cubic	
Trigonometric	
Reciprocal	

[2 marks]

Exam-style questions

13.4 Sketch the curve with equation $y = 10^x$

Give the coordinates of any points of intersection with the axes.

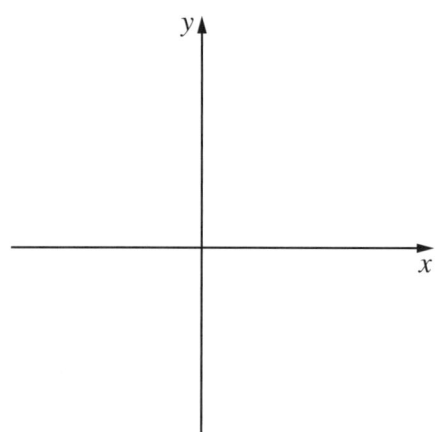

[3 marks]

13.5 The graph of the curve C with equation $y = f(x)$ is transformed to give the graph of the curve T with equation $2f(x - 2)$.

The point on C with coordinates $(3, 5)$ is mapped to point P on T.

Find the coordinates of P. **[2 marks]**

13.6 The graph of $y = f(x)$ is shown below. On the same set of axes, sketch the graph of $y = f(2x)$.

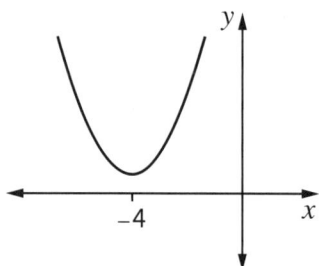

[2 marks]

13.7 Graph A has been transformed to give graph B.

The equation of graph A is $y = f(x)$

Write down the equation of graph B.

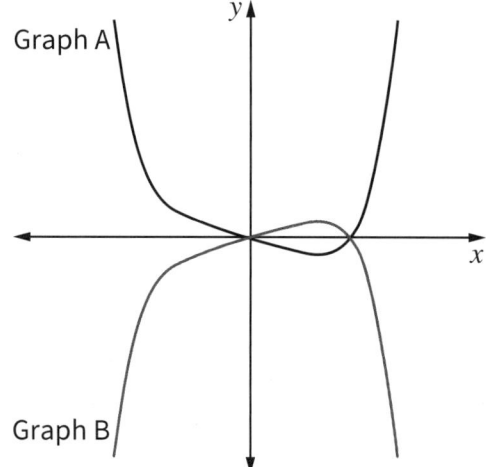

[1 mark]

13.8 Sketch the graph of $y = 0.5^x$ for $-2 \leq x \leq 3$

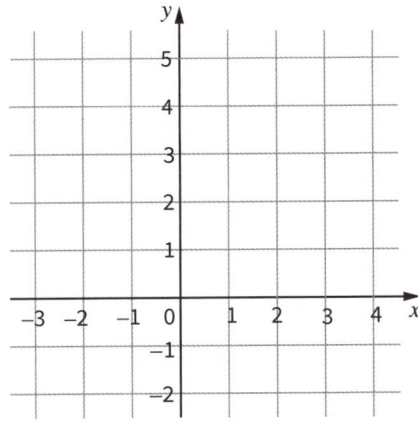

[3 marks]

13.9 The graph with equation $y = x^2 - 4x + 1$ is shown.

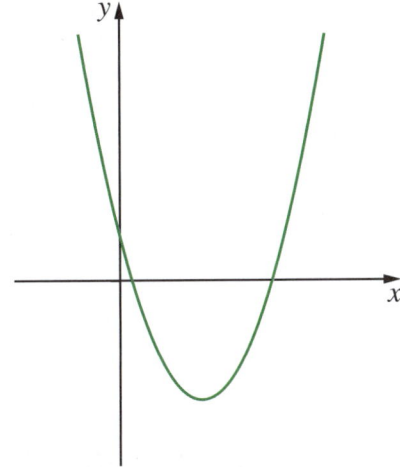

(a) Give the coordinates of the minimum point of the graph with equation $y = x^2 - 4x + 1$ [2 marks]

(b) On the same axes, sketch the graph with equation $y = (x + 1)^2 - 4(x + 1) + 1$ and give the coordinates of the minimum point. [1 mark]

Questions referring to previous content

13.10 Solve algebraically the pair of simultaneous equations.

$5x + y = 10 \qquad 10xy = -48$ [5 marks]

13.11 The first term of an arithmetic sequence is 4 and the fourth term is 25. Find the common difference of the sequence and determine if 61 is a term in the sequence. [4 marks]

Knowledge

14 Non-linear real-life graphs

Gradients on non-linear real-life graphs

To calculate the average **gradient** between two points:
- draw a **chord** connecting the two points
- calculate the gradient of the chord.

To estimate the gradient at a point:
- draw a **tangent** to the curve at that point
- calculate the gradient of the tangent.

For a distance–time graph:
- the gradient of the chord gives average speed between the two points
- the gradient of the tangent gives an estimate of the speed at that point.

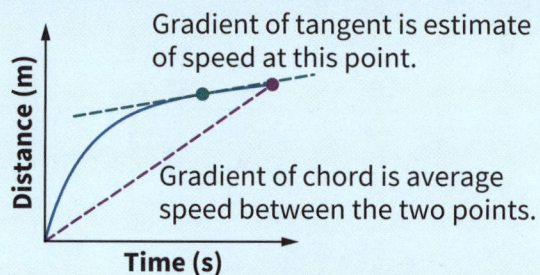

Gradient of tangent is estimate of speed at this point.

Gradient of chord is average speed between the two points.

LINK
You will revise tangents and chords in Chapter 19.

Worked example

The distance–time graph for a train on part of its journey is shown.

(a) Calculate the train's average speed in km/h during the first four minutes.

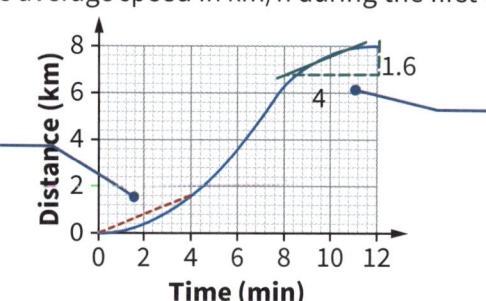

Draw a chord connecting the points on the curve where $x = 0$ and $x = 4$

Draw a tangent to the graph at $x = 10$ and a triangle to work out the gradient.

gradient of chord $= \dfrac{\text{change in } y}{\text{change in } x}$ — Find the gradient of the chord.

$= \dfrac{1.6}{4}$

$= 0.4$ — Convert the speed to km/h.

average speed $= 0.4\,\text{km/min}$

$= 24\,\text{km/h}$

WATCH OUT
Pay attention to the scale of the axes!

(b) Estimate the train's speed after 10 minutes in kilometres per hour.

gradient of tangent $= \dfrac{1.6}{4} = 0.4$ — Calculate the gradient of the tangent to find the speed in km/min.

speed $= 24\,\text{km/h}$

Area under the curve on non-linear real-life graphs

For a velocity–time graph:
- the gradient of the tangent gives an estimate of the acceleration at that point
- the area under the graph gives the distance travelled up to that point.

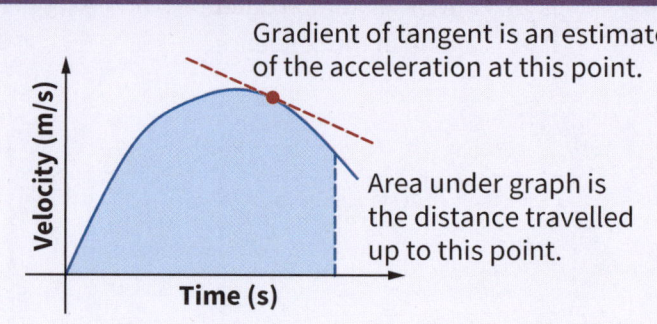

Gradient of tangent is an estimate of the acceleration at this point.

Area under graph is the distance travelled up to this point.

14

Area under the curve on non-linear real-life graphs

When estimating the area under a curve, the answer could be an over-estimate, or an under-estimate.

Your answer could either be an under-estimate or an over-estimate

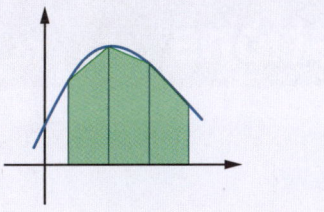 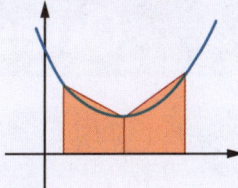

of the actual area.

Worked example

Here is the velocity–time graph for an object in motion.

(a) Estimate the distance travelled by the object in the first 12 s.

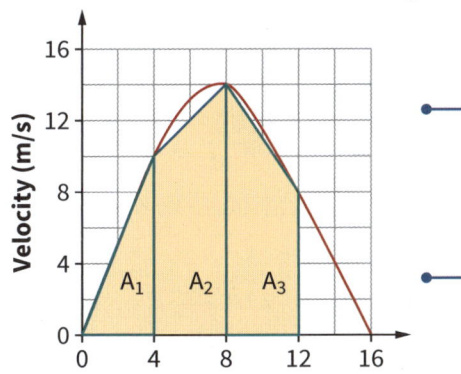

Divide the area into three strips of equal width by drawing vertical lines at 4 s, 8 s, and 12 s.

Draw a chord between the two outer points of each strip to approximate the curve.

$A_1 = \frac{1}{2} \times (10 + 0) \times 4 = 20$

Calculate each area of each of the trapeziums.

$A_2 = \frac{1}{2} \times (10 + 14) \times 4 = 48$

$A_3 = \frac{1}{2} \times (14 + 8) \times 4 = 44$

$20 + 48 + 44 = 112$, so the object travelled 112 m.

Add the areas together to find the distance travelled.

(b) Is your answer to part **(a)** an overestimate or an underestimate of the actual distance? Give a reason for your answer.

It is an underestimate since all the chords lie under the curve.

Key terms — Make sure you can write a definition for these key terms: chord, gradient, tangent

14 Knowledge

Retrieval

14 Non-linear real-life graphs

Learn the answers to the questions below, then cover the answers column with a piece of paper and write as many as you can. Check and repeat.

	Questions	Answers
1	For a distance–time graph, what does the gradient represent?	Speed.
2	When you find the area under a curve, why is it an estimate?	You use trapeziums rather than following the exact curve.
3	When is your answer an underestimate?	When the tops of the trapeziums are below the curve.
4	On a velocity–time graph, what does the area under the curve represent?	Distance.
5	How do you calculate the gradient between 2 points?	Draw a chord and calculate the gradient.
6	How do you estimate the gradient at a point?	Draw a tangent and calculate the gradient.
7	How do you calculate gradient?	Change in y over change in x.
8	On a velocity time curve, what does the gradient of a tangent tell us?	The acceleration at that point.
9	What is the formula for the area of a trapezium?	$\frac{1}{2}(a+b)h$

Previous questions

Now go back and use these questions to check your knowledge of previous topics.

	Questions	Answers
1	How do you simplify a surd?	Look for a factor that is a square number.
2	What is a linear graph?	A linear graph is a straight line which can be horizontal, vertical or sloping.
3	What are the two algebraic methods you can use to solve simultaneous equations?	Elimination and substitution.

Practice 14

Exam-style questions

14.1 The velocity–time graph shows the velocity, v m/s, of a particle t seconds after it is released.

(a) Using two strips of equal width, estimate the area under the graph. [3 marks]

(b) Using four strips of equal width, estimate the area under the graph. [3 marks]

(c) Explain why your answer to part **(a)** is smaller than your answer to part **(b)**. [1 mark]

(d) Explain what the area under the graph represents. [1 mark]

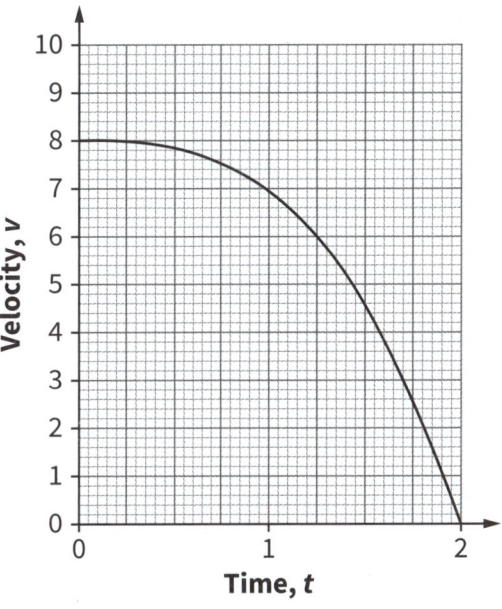

14.2 The speed–time graph shows the speed, v m/s, of a ball as it rolls along a road for t seconds.

(a) Using three strips of equal length, estimate the total distance the ball travels in the first three seconds. [3 marks]

(b) (i) Using three strips of equal length, estimate the total distance the ball travels in the second three seconds. [3 marks]

   (ii) State whether your answer to part **(b)(i)** is an overestimate or an underestimate and give a reason. [1 mark]

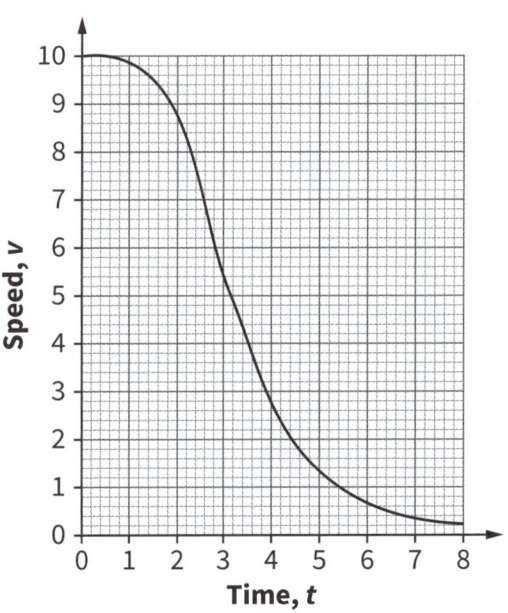

Exam-style questions

14.3 The graph shows the velocity, v, of a skateboarder over a period of time, t.

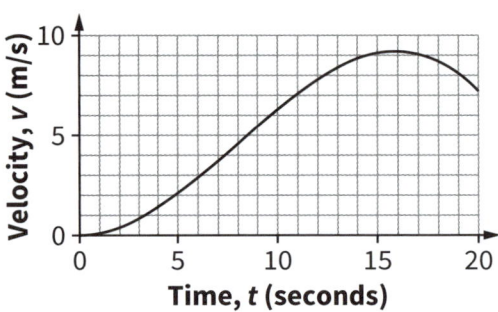

(a) Estimate the acceleration when $t = 5$ [3 marks]

(b) (i) Using two strips of equal width, estimate the area under the curve between $t = 15$ and $t = 20$ [3 marks]

Show all working and give your answer to 2 s.f.

(ii) State what your answer to part (i) represents. [1 mark]

14.4 David mows his lawn every few weeks. The graph shows the height of his lawn over a period of 12 weeks one summer.

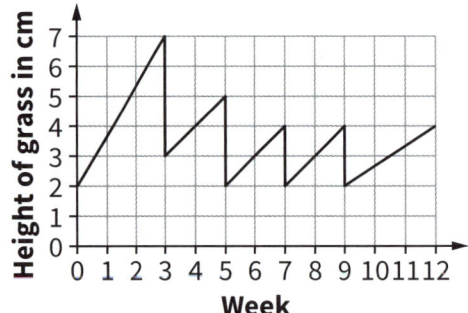

(a) In which weeks does David cut the lawn? [3 marks]

(b) Between which weeks does the grass grow at the slowest rate? [1 mark]

(c) What was the rate of growth of the grass, in cm/week, between weeks 5 and 7? [1 mark]

14.5 Here are the cross-sections of three swimming pools.

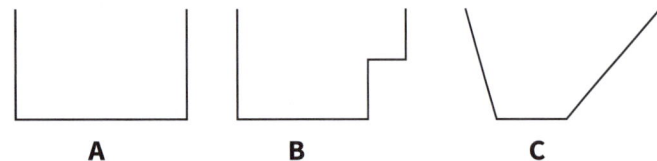

Each pool is filled with water from a hose at the same rate.

Match each pool to the graph of depth of water in the pool against time.

> **EXAM TIP**
> The steeper the graph, the faster the rate of change.

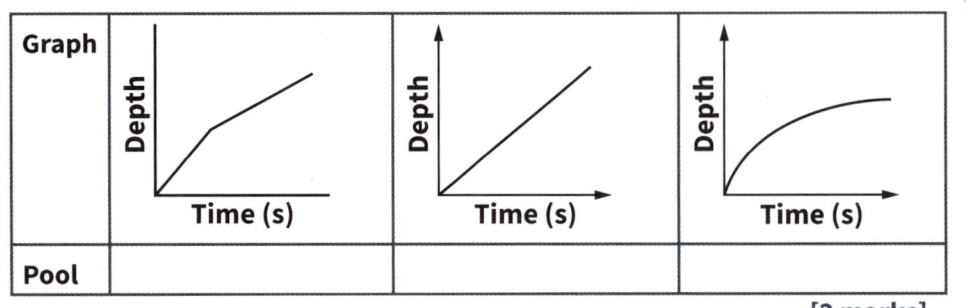

[2 marks]

14.6 A ball is released from the top of a tower. The distance–time graph shows information about its fall to the ground.

(a) Calculate the average speed of the ball in the first two seconds. **[2 marks]**

(b) Use the graph to estimate the speed of the ball at time 1 second. **[3 marks]**

EXAM TIP
To find the average speed, draw a chord. To find the speed at a certain point, draw a tangent.

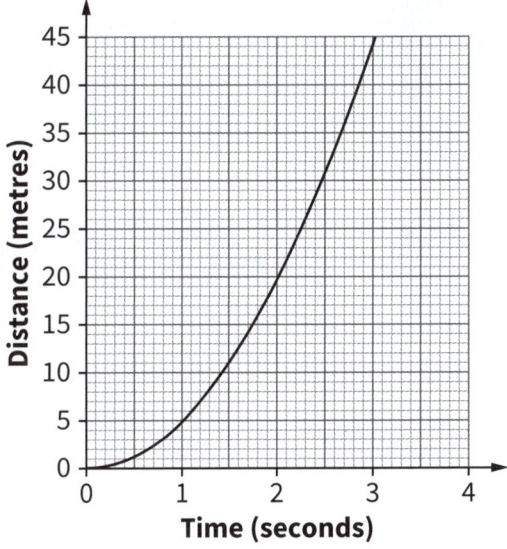

14.7 Janusz goes on a train journey. The speed–time graph shows the speed of the train throughout the journey.

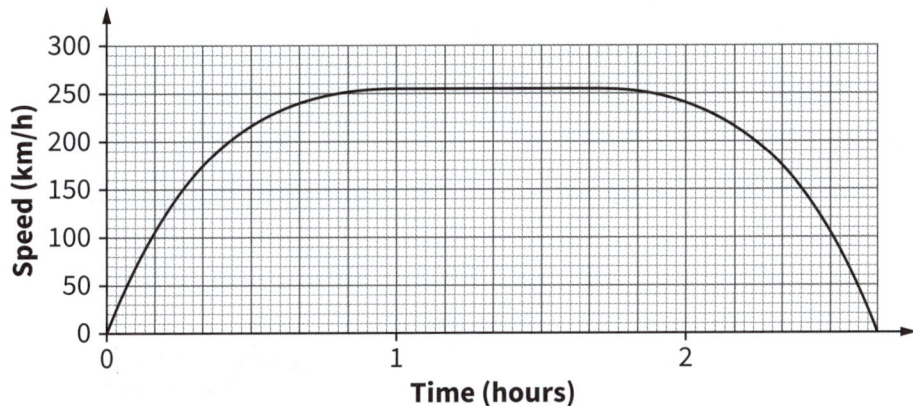

(a) Calculate the train's average acceleration during the first 30 minutes of the journey. **[2 marks]**

(b) Explain what is happening to the train's speed after 1.75 hours of travel. **[1 mark]**

(c) Calculate the train's deceleration at 2 hours. **[3 marks]**

Questions referring to previous content

14.8 Fill in the blanks in these sentences:

All exponential graphs of the form where $y = k^x$ is a positive constant, pass through the point with coordinates _____ . When $k > 1$ the graph will demonstrate exponential _____, and when $k < 1$ it demonstrates exponential _____ . **[3 marks]**

14.9 The function $y = f(x)$ undergoes a transformation given by $y = f(x - 2)$. Identify where the point (7, 14) on the curve $y = f(x)$ will be when translated by $y = f(x - 2)$. **[1 mark]**

14 Practice 107

Knowledge

15 Algebraic fractions, rearranging formulae with algebraic fractions, proof, functions and composite functions

Algebraic fractions

An **algebraic fraction** is a fraction that contains at least one variable.

These are all examples of algebraic fractions: $\dfrac{3ab}{c^2} \quad \dfrac{x}{2} \quad \dfrac{3x^2 + 2x + 1}{5x - 4}$

Algebraic fractions can be simplified by cancelling common factors in the numerator and denominator.

Worked example

1. Simplify $\dfrac{24a}{36a^3}$

   $= \dfrac{24a}{36a^3}$ — Simplify the multipliers by dividing by the HCF.

   $= \dfrac{2a}{3a^3}$ — Simplify the 'a's by cancelling the indices.

   $= \dfrac{2}{3a^2}$

   LINK
   To remind yourself about the rules of dividing indices, refer to Chapter 5.

2. Simplify $\dfrac{(b-1)(b+3)}{2(b-1)^2}$

   $\dfrac{(b-1)(b+3)}{2(b-1)^2}$ — Expand the squared term (indices before multiplication).

   $= \dfrac{\cancel{(b-1)}(b+3)}{2(b-1)\cancel{(b-1)}}$ — Cancel the factors that appear in both the numerator and denominator.

   $= \dfrac{(b+3)}{2(b-1)}$ — Write the simplified fraction.

3. (a) Simplify $\dfrac{x^2 + 3x - 10}{x^2 - 25}$

   $\dfrac{x^2 + 3x - 10}{x^2 - 25}$ — Factorise the numerator and denominator.

   $= \dfrac{(x+5)(x-2)}{(x+5)(x-5)}$

   $= \dfrac{(x-2)}{(x-5)}$ — Cancel the common factor (terms that appear in both the numerator and denominator).

   WATCH OUT
   The remaining fraction will not simplify any further since there are no common factors of the numerator and denominator. Be careful, x is not a common factor here since it isn't a multiplier, so you can't cancel it.

   (b) Hence, solve $\dfrac{x^2 + 3x - 10}{x^2 - 25} = 2$

   $\dfrac{(x-2)}{(x-5)} = 2$ — Replace the LHS with the simplified equation from part **(a)**.

   $x - 2 = 2(x - 5)$

   $x - 2 = 2x - 10$ — Rearrange the equation. Expand the brackets. Collect like terms to solve.

   $x = 8$

15

Multiplying algebraic fractions

To multiply algebraic fractions

- multiply the numerators together
- multiply the denominators together
- then cancel the common factors.

To divide by an algebraic fraction, multiply by its **reciprocal**.

LINK

To remind yourself about calculating with fractions, look back at Chapter 4.

Worked example

Simplify fully:

(a) $\dfrac{x^2 + 6x}{x^2 + 3x + 2} \times \dfrac{x^2 - 4x - 5}{3x^2}$

$= \dfrac{x(x + 6)}{(x + 1)(x + 2)} \times \dfrac{(x + 1)(x - 5)}{3x^2}$

Factorise the numerator and denominator of both fractions where possible.

$= \dfrac{x(x + 6)(x + 1)(x - 5)}{3x^2(x + 1)(x + 2)}$

Multiply the two numerators and the two denominators.

$= \dfrac{(x + 6)(x - 5)}{3x(x + 2)}$

Simplify by cancelling common factors.

WATCH OUT

Do not expand brackets when multiplying algebraic fractions.

(b) $\dfrac{5x + 10}{x^2 - 9} \div \dfrac{5x^2}{x + 3}$

$= \dfrac{5x + 10}{x^2 - 9} \times \dfrac{x + 3}{5x^2}$

Rewrite the calculation as a multiplication, swapping the numerator and the denominator of the second fraction.

$= \dfrac{5(x + 2)}{(x - 3)(x + 3)} \times \dfrac{x + 3}{5x^2}$

Factorise where possible.

$= \dfrac{5(x + 2)\cancel{(x + 3)}}{5x^2(x - 3)\cancel{(x + 3)}}$

Multiply the fractions together.

WATCH OUT

You can cancel before or after multiplying, but be careful to only cancel common factors of a numerator and a denominator.

$= \dfrac{x + 2}{x^2(x - 3)}$

Cancel common factors.

Knowledge

15 Algebraic fractions, rearranging formulae with algebraic fractions, proof, functions and composite functions

Adding or subtracting algebraic fractions

Algebraic fractions can be added or subtracted.

Worked example

Simplify fully $\dfrac{x+1}{x+3} + \dfrac{x-1}{x^2+4x+3}$

$= \dfrac{x+1}{x+3} + \dfrac{x-1}{(x+1)(x+3)}$ — Factorise where possible.

$= \dfrac{(x+1)(x+1)}{(x+1)(x+3)} + \dfrac{x-1}{(x+1)(x+3)}$ — Find the **LCM** of the denominators and rewrite as equivalent fractions. Remember to multiply the numerator by the same factor as the denominator.

$= \dfrac{(x+1)(x+1) + (x-1)}{(x+1)(x+3)}$

$= \dfrac{x^2 + 2x + 1 + x - 1}{(x+1)(x+3)}$ — Add the fractions.

$= \dfrac{x^2 + 3x}{(x+1)(x+3)}$ — Expand the brackets in the numerator.

$= \dfrac{x(x+3)}{(x+1)(x+3)}$ — Collect like terms to simplify.

$= \dfrac{x}{x+1}$ — Factorise the numerator.

— Cancel common factors.

WATCH OUT

Always factorise and find the LCM of the denominators. If you cross-multiply by the denominators without factorising and using the LCM, the algebra can get quite tricky.

Proof

An **equation** is only true for certain values.
For example, $x + 5 = 8$ is only true for $x = 3$.

An **identity** is true for all values.
For example, $2x + 3x = 5x$ is true for all values of x, so you can write $2x + 3x \equiv 5x$.

If you are asked to **prove** a result, it will always be an identity.

To prove an identity, start with the expression on one side and use algebraic techniques, such as expanding and simplifying, until you reach the expression on the other side.

REVISION TIP

Equation = for certain values

identity $\equiv$ for all values

Worked example

1. Prove that $(a+b)^2 - (a-b)^2 \equiv 4ab$

   LHS $\equiv (a+b)^2 - (a-b)^2$ — Start with the left-hand side (LHS). Expand the brackets.

   $\equiv (a^2 + 2ab + b^2) - (a^2 - 2ab + b^2)$

   $\equiv a^2 + 2ab + b^2 - a^2 + 2ab - b^2$ — Collect like terms.

   $\equiv 4ab$

   $\equiv$ RHS — Compare to the right-hand side (RHS).

Proof

> **Worked example**
>
> Prove that the difference between the squares of any two consecutive odd numbers is always a multiple of 8.
>
> Let n be an integer. — *If n is an integer, you can use $2n$ to represent even numbers and $2n + 1$ to represent odd numbers.*
>
> Then $2n + 1$ and $2n + 3$ are consecutive odd numbers. — *Write an expression to represent the statement.*
>
> difference between the squares $= (2n + 3)^2 - (2n + 1)^2$
>
> — *Expand the brackets.*
>
> $= (4n^2 + 12n + 9) - (4n^2 + 4n + 1)$
>
> — *Collect like terms to simplify.*
>
> $= 8n + 8$
>
> $= 8(n + 1)$ — *To show an expression is a multiple of a number, write it with the number as a factor.*

If a result is *not* true, you only need to find one example to disprove the result. This is called a counterexample.

> **Worked example**
>
> Anis claims that the difference between two square numbers is always even. Give a counterexample to disprove her claim.
>
> $2^2 = 4$ and $3^2 = 9$
>
> $9 - 4 = 5$, which is an odd number. This disproves Anis's claim.

Functions

Functions can be labelled as f(x), g(x), and so on. For example, f(x) = x^2.

You can **evaluate** a function by substituting a value in place of x.

> **Worked example**
>
> f(x) = $4x^2 + 2x - 1$
>
> **(a)** Find f(−3)
>
> f(−3) = $4(-3)^2 + 2(-3) - 1$ — *Substitute −3 for x.*
>
> $= 36 - 6 - 1$ — *Expand the brackets and simplify.*
>
> $= 29$
>
> **(b)** Find values of x for which f(x) = 1
>
> $4x^2 + 2x - 1 = 1$ — *Set the function equal to 1*
>
> $4x^2 + 2x - 2 = 0$ — *Subtract 1 from both sides to get a quadratic equal to 0 and solve.*
>
> $2x^2 + x - 1 = 0$
>
> $(2x - 1)(x + 1) = 0$ — *Factorise the equation.*
>
> $x = \frac{1}{2}$ or $x = -1$ — *Then set each factor equal to 0 to find the two values of x.*

LINK
To remind yourself about the order of operations, see Chapter 1.

LINK
To remind yourself about solving quadratic equations, see Chapter 9.

Knowledge

15 Algebraic fractions, rearranging formulae with algebraic fractions, proof, functions and composite functions

Composite functions

Functions can be combined to form **composite functions**, such as fg(x).

The function closest to (x) is carried out first.

For example, if $f(x) = x^2$ and $g(x) = x + 2$

fg(x) = f(x + 2) ← fg(x) means function g followed by function f.
    = $(x + 2)^2$

gf(x) = g(x^2) ← gf(x) means function f followed by function g.
    = $x^2 + 2$

Worked example

$f(x) = 2x + 1$, $g(x) = 5 - x^2$

(a) Work out gf(2).

f(2) = 2(2) + 1 = 5 ← Substitute 2 into f(x).

gf(2) = g(5)

= $5 - 5^2$ ← Substitute the answer into g(x).

= −20

(b) Write a simplified expression for fg(x).

fg(x) = f(5 − x^2) ← Substitute the expression for g.

= 2(5 − x^2) + 1 ← Substitute the expression for x.

= 11 − $2x^2$ ← Expand and simplify.

(c) Solve the equation fg(x) = g(x).

11 − $2x^2$ = 5 − x^2 ← Write the substituted expressions as equal.

$x^2 = 6$ ← Collect like terms.

$x = \sqrt{6}$ or $x = -\sqrt{6}$ ← Rearrange to solve.

Inverse functions

The **inverse** of a function reverses its effect.

The inverse of f(x) is written $f^{-1}(x)$.

The inverse of a function can be found using a function machine. For example, the function machine for $f(x) = 3x - 7$ is:

$x \rightarrow$ × 3 $\rightarrow$ − 7 $\rightarrow$ f(x)

To find $f^{-1}(x)$, replace each operation with its inverse and work from right to left:

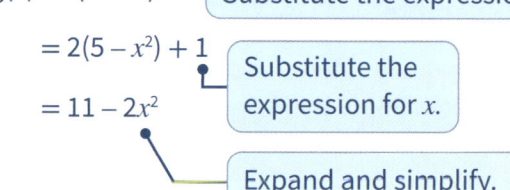

$f^{-1}(x) = \dfrac{x + 7}{3}$

Inverse functions

So $f^{-1}(x) = \dfrac{x+7}{3}$

The inverse can also be found algebraically by rearranging the formula to make x the **subject**.

Worked example

The function f is such that $f(x) = 9 - 5x$.

(a) Find $f^{-1}(x)$.

$y = 9 - 5x$ — Write the formula, replacing $f(x)$ with y.
$5x = 9 - y$
$x = \dfrac{9-y}{5}$ — Rearrange the formula to make x the subject.
$f^{-1}(x) = \dfrac{9-x}{5}$

Replace:
- x with $f^{-1}(x)$
- y with x.

(b) Solve the equation $f(x) = f^{-1}(x)$.

$9 - 5x = \dfrac{9-x}{5}$ — Write an equation using the expressions for $f(x)$ and $f^{-1}(x)$.
$45 - 25x = 9 - x$
$36 = 24x$
$x = \dfrac{36}{24}$ — Rearrange and collect like terms to solve.
$= 1.5$

Worked example

$f(x) = 5x - 2$, $g(x) = x^3 + 1$

The function h is such that $h(x) \equiv gf(x)$.

(a) Find $h^{-1}(x)$.

$h(x) = g(5x - 2)$ — Write $h(x)$ using the functions given.
$= (5x - 2)^3 + 1$
$y = (5x - 2)^3 + 1$
$y - 1 = (5x - 2)^3$ — Rearrange to find the inverse.
$\sqrt[3]{y - 1} = 5x - 2$
$2 + \sqrt[3]{y - 1} = 5x$
$x = \tfrac{1}{5}\left(2 + \sqrt[3]{y - 1}\right)$
$h^{-1}(x) = \tfrac{1}{5}\left(2 + \sqrt[3]{y - 1}\right)$

Replace x with $h^{-1}(x)$, and replace y with x.

(b) Given that $h(a) = 9$, find the value of a.

If $h(a) = 9$ — Substitute the value into the equation and solve.
$a = h^{-1}(9)$
$= \tfrac{1}{5}\left(2 + \sqrt[3]{9 - 1}\right)$
$= 0.8$

Key terms — Make sure you can write a definition for these key terms.

algebraic fraction   composite function   equation   evaluate
inverse   inverse function   identity   lowest common multiple
proof   reciprocal   subject

Retrieval

15 Algebraic fractions, rearranging formulae with algebraic fractions, proof, functions and composite functions

Learn the answers to the questions below, then cover the answers column with a piece of paper and write as many as you can. Check and repeat.

Questions | Answers

#	Question	Answer
1	How do you simplify algebraic fractions?	Look for common factors to cancel – you may need to factorise the numerator and/or the denominator first.
2	How do you add or subtract algebraic fractions?	Find a common denominator. You may need to factorise first.
3	What does 'make x the subject of the equation' mean?	Rearrange the equation into the form $x = ...$
4	What symbol does an identity include?	The identity symbol, $\equiv$.
5	If n is an integer, how could you represent an even number?	$2n$
6	If n is an integer, how could you represent an odd number?	$2n + 1$ or $2n - 1$ (other answers are possible).
7	How do you represent a function?	$f(x)$ or $g(x)$ (other letters can be used).
8	What is a composite function?	Two or more functions combined.
9	In $fg(x)$, which function happens first?	$g(x)$
10	What are the three steps you need to follow to find the inverse of a function $f(x)$?	1. Write in the form $y = ...$ 2. Rearrange to make x the subject. 3. Replace y with x and x with y, and write as $f^{-1}(x) = ...$

Previous questions
Now go back and use these questions to check your knowledge of previous topics.

Questions | Answers

#	Question	Answer
1	What is an integer?	A whole number.
2	What is a surd?	An irrational square root.
3	What is an expression?	A collection of letters and numbers which cannot be solved (no equals or inequality sign).
4	How do you solve two inequalities that require you to show the solutions on a number line?	Solve each inequality separately then combine the solutions.
5	What are the two algebraic methods you can use to solve simultaneous equations?	Elimination and substitution.

Practice

Exam-style questions

15.1 **(a)** Expand fully $(x-1)(x-2)(x-3)$. [3 marks]

(b) Using part **(a)**, prove that $n^4 - 6n^3 + 11n^2 - 6n$ is always even for all integer values of $n > 3$. [2 marks]

15.2 Simplify

(a) $\dfrac{4x^2 - 12x}{2x}$ [2 marks]

(b) $\dfrac{x^2 - x - 2}{x^2 - 6x + 8}$ [3 marks]

15.3 Simplify [3 marks]

(a) $\dfrac{3x - 4}{6x^2 + 7x - 20}$

(b) Hence, solve the equation $\dfrac{3x - 4}{6x^2 + 7x - 20} = 1$ [2 marks]

> **EXAM TIP**
> Replace the fraction in part **b** with its simplified form from part **a**.

15.4 Solve the equation $\dfrac{1}{x-3} = \dfrac{x}{x+5}$ [3 marks]

15.5 Simplify $\dfrac{2x - 8}{3x - 15} \times \dfrac{x - 5}{x - 3}$ [3 marks]

15.6 Simplify $\dfrac{x}{x^2 + 2x - 35} \div \dfrac{6x^3}{3x + 21}$ [4 marks]

15.7 Given that $x^2 : (x-1) = 4 : 1$, find the value of x. [4 marks]

15.8 Write each of these as a single fraction in its simplest form.

(a) $\dfrac{4}{x+1} + \dfrac{3}{x+2}$ [3 marks]

(b) $\dfrac{x}{x-2} - \dfrac{x-2}{x+3}$ [3 marks]

Exam-style questions

15.9 An equilateral triangle and an isosceles triangle have side lengths as shown in the diagram. Find the difference between the perimeters of the two triangles.

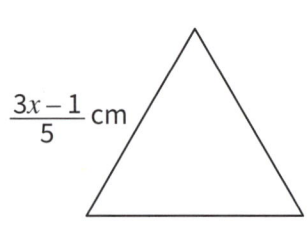
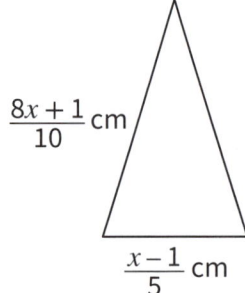

[4 marks]

15.10 $f(x) = x^2 - 2x$

   (a) Find the value of

      (i) $f(1)$ [2 marks]

      (ii) $f(-5)$.   [2 marks]

   (b) Find the values of x for which

      (i) $f(x) = 15$ [3 marks]

      (ii) $f(x) = 4x$.   [3 marks]

15.11 $f(x) = 3x - 1$ and $g(x) = \dfrac{1}{x}$

   Find the value of

   (a) $gf(1)$ [3 marks]

   (b) $fg\left(\dfrac{1}{2}\right)$ [3 marks]

15.12 Find the inverse of these functions.

   (a) $f(x) = \dfrac{x-1}{2}$ [3 marks]

   (b) $g(x) = x^2 - 4, x \geq 0$ [3 marks]

   (c) $h(x) = \dfrac{x}{2x+3}, x \neq -\dfrac{3}{2}$ [4 marks]

15.13 Prove that $(m+n)^2 + (m-n)^2 \equiv 2(m^2 + n^2)$.   [3 marks]

15.14 Prove algebraically that these statements are true.

   (a) The sum of five consecutive integers is divisible by 5.   [3 marks]

   (b) The sum of the squares of any two even numbers is divisible by 4.   [3 marks]

> **EXAM TIP**
> Use two different letters to represent the two numbers.

15.15 Show that $\dfrac{2x^2 + 3x}{9x^2 - 4} \times \dfrac{9x^2 + 42x + 24}{x^2 + 4x}$ simplifies to $\dfrac{a(bx + c)}{dx + e}$,

where a, b, c, d and e are integers. [4 marks]

15.16 Express $\dfrac{2}{x - 1} + \dfrac{3}{x - 2} + \dfrac{3}{2x}$ as a single fraction in its simplest form. [4 marks]

15.17 $f(x) = \dfrac{1}{x}$ $g(x) = \dfrac{2}{x^2}$

(a) Find fg(1). [2 marks]

(b) Find gf(x). [2 marks]

15.18 $f(x) = 2x + 1$ $g(x) = x^2$

(a) Find ff(x). [2 marks]

(b) Given that $f^{-1}(x) = fg(x)$, show that $4x^2 - x + 3 = 0$. [6 marks]

15.19 Prove that the sum of two consecutive odd numbers is always a multiple of 4. [3 marks]

Questions referring to previous content

15.20 Multiply these numbers and write your answer in standard form:

$27.2 \times 10^6 \times 1.75 \times 10^{-3}$ [1 mark]

15.21 Factorise this expression: $-12xy^2 + 15xyz - 3x^2y + 21ax^3y^2$ [2 marks]

Knowledge

16 Ratio

Ratios: an overview

A **ratio** describes how things are split into **parts**.

The ratio of black beads to red beads on this bracelet is 3:2.

Worked example

Simplify the ratio 49:28

$\left(\frac{49}{7}\right):\left(\frac{28}{7}\right) = 7:4$

The HCF of 49 and 28 is 7, so divide both by 7.

Ratio problems can sometimes be solved by 'scaling up' a ratio. You do this by multiplying both sides by the same number.

LINK
To remind yourself about HCF, look back at Chapter 3.

Ratios compare parts to parts; **fractions** compare parts to the whole. For example, if the ratio of adults to children is 1:2 then $\frac{1}{3}$ of the total are adults and $\frac{2}{3}$ are children.

Ratios can also have three or more parts, for example $a:b:c$.

A ratio can be simplified fully by dividing both sides by their highest common factor (HCF).

Using ratio

WATCH OUT

These next two questions look very similar. Make sure you focus on the **information** and **instructions** given.

Worked example

Ben, Sima, and Lisa share some money in the ratio 3:4:5. Lisa gets £30. How much does Sima get?

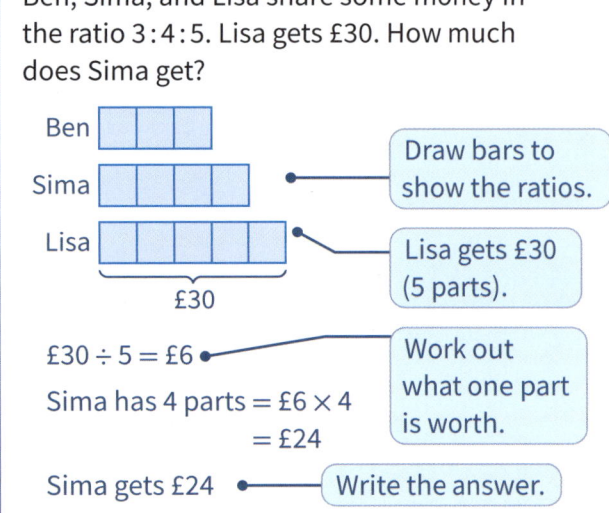

Draw bars to show the ratios.

Lisa gets £30 (5 parts).

£30 ÷ 5 = £6

Sima has 4 parts = £6 × 4 = £24

Work out what one part is worth.

Sima gets £24

Write the answer.

Worked example

Ben, Sima, and Lisa share some money in the ratio 3:4:5. Lisa gets £30 more than Ben. How much does Sima get?

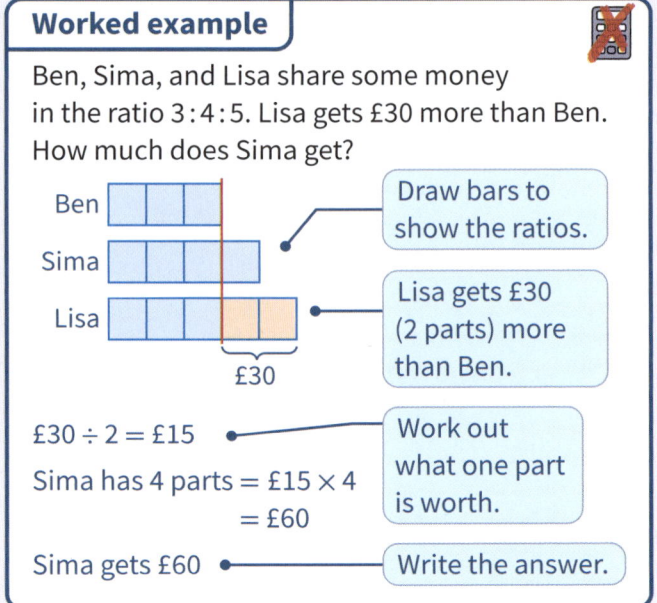

Draw bars to show the ratios.

Lisa gets £30 (2 parts) more than Ben.

£30 ÷ 2 = £15

Sima has 4 parts = £15 × 4 = £60

Work out what one part is worth.

Sima gets £60

Write the answer.

16

Using ratio

Worked example

In a class of children, the ratio of 8-year-olds to 9-year-olds is 2:3.

25% of the 8-year-olds are left-handed.

$\frac{1}{3}$ of the 9-year-olds are left-handed.

What percentage of the class is right-handed?

8-year-olds [|] ← Draw bars to show the ratios.

9-year-olds [| |]

8-year-olds [L | | |] ← Split the 8-year-olds boxes into 4 so you can show 25% are left-handed. Mark $\frac{1}{3}$ of the 9-year-olds boxes to show left-handed.

9-year-olds [L | | | |]

8-year-olds [L | R | R | R] ← Make all the boxes the same size. Mark and count the right-handed boxes.

9-year-olds [L | L | R | R | R | R]

$\frac{7}{10}$ are right-handed = 70% ← Write the answer.

Ratios and scale

A **scale** can be written as a ratio, such as 1:400, or as a sentence.

You multiply the length on the drawing to get the real-life length.

You divide the real-life length to get the length on the drawing.

Worked example

A drawing of a tower is 1.4 cm tall. It is drawn to a scale of 1:400

How tall is the tower in real life?

drawing	real life
1	400
1 cm	400 cm
1 cm	4 m

← Make a table.
← Write the scale.
← Put in the units.
← Change to better units.

Size of drawing = 1.4 cm

Actual size = 1.4 × 4
            = 5.6 m

Worked example

On a map, 2 cm represents 3 km.

Write the scale of the map as a ratio in the form 1:n

map	real
2 cm	3 km
2 cm	3000 m
2 c̶m̶	300 000 c̶m̶
1	150 000

← Simplify.

16 Knowledge 119

Knowledge

16 Ratio

Problem solving with ratios

Worked example

Oskar mixes red and yellow paint in the ratio 2 : 3

1. Oskar needs 240 ml of red paint to paint a room.

   How much yellow paint does he need?

   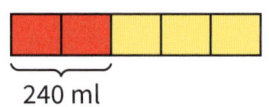
   240 ml

   — Draw a **bar model** in the ratio 2 : 3

   — 240 ml is 2 parts so 1 part
   $= 240 \div 2 = 120$
   Yellow paint is 3 parts, so 3 parts
   $= 120 \times 3 = 360$

   Oskar needs 360 ml of yellow paint.

2. Oskar needs 15 litres of the mixed paint to paint his house.

   How much red paint does he need?

   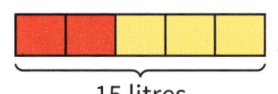
   15 litres

   — 1 part $= 15 \div 5 = 3$
   Red $= 2$ parts $= 3 \times 2 = 6$

   Oskar needs 6 litres of red paint.

Using fractions and percentages with ratio

If $a : b = 3 : 5$, then $\frac{a}{b} = \frac{3}{5}$

If b is 20% less than a, then $a : b = 100 : 80$, which is 5 : 4.

If $2a = 3b$, then $a : b = \frac{3}{2} : 1$, which is 3 : 2 (be careful here: a is greater than b).

Worked example

A bag contains green and blue beads in the ratio 3 : 4.

Write an equation for y in terms of x to show the relationship between the number of green beads (y) and the number of blue beads (x).

$\frac{3}{4} : 1$ — Write the ratio in the form $n : 1$
Turn this into an equation.

$y = \frac{3}{4}x$ — You can substitute in some values to check.

Using fractions and percentages with ratio

In questions where the ratio changes:

- Write an equation for each situation using fractions.
- Form linear equations.
- Solve the equations simultaneously.

LINK
To remind yourself about solving simultaneous equations, look back at Chapter 11.

Worked example

There are red and blue counters in a bag in the ratio 1 : 5.

Three red counters are added to the bag. The ratio of red to blue counters is now 1 : 4.

How many blue counters are there in the bag?

Let x be the number of red counters at the start.

Write an equation for each situation.

At the start, there are:

- x red counters
- $5x$ blue counters
- $6x$ counters altogether

Let y be the number of red counters at the end.

At the end, there are:

- y red counters
- $4y$ blue counters
- $5y$ counters altogether

3 red counters are added to the bag, so

$y = x + 3$ (equation 1)

Total increase in counters is 3, so

$5y = 6x + 3$ (equation 2)

Write the equations as simultaneous equations.

$5(x + 3) = 6x + 3$

$5x + 15 = 6x + 3$

Expand the brackets and solve to find x.

$15 - 3 = 6x - 5x$

$12 = x$

$y = x + 3$

$y = 12 + 3$

Use substitution to find y.

$y = 15$

blue counters at end $= 4y$

$= 4 \times 15$

Multiply to find number of blue counters at the end.

$= 60$ blue counters.

Key terms — Make sure you can write a definition for these key terms: bar model, fraction, parts, ratio, scale

Retrieval

16 Ratio

Learn the answers to the questions below, then cover the answers column with a piece of paper and write as many as you can. Check and repeat.

	Questions	Answers
1	What is a ratio used for?	To compare two or more quantities in relation to each other.
2	How do we write a ratio in the form $1 : n$?	Divide both values to make the first part 1.
3	How do we know when a ratio is written in its simplest form?	The only remaining common factor of each of the parts of the ratio is 1.
4	To simplify a ratio fully, what do you need to divide all the parts in the ratio by?	The highest common factor of the parts.
5	The ratio of adults to children in a group is $4 : 7$. Explain how to express the number of adults as a fraction of the whole group.	There are 4 parts representing adults and $4 + 7 = 11$ parts in total, so the fraction of adults is $\frac{4}{11}$.
6	How do you scale up a ratio?	Multiply each part by the same number.
7	True or false? A ratio can only have two parts.	False, ratios can have more than two parts, e.g. $3 : 4 : 6$.
8	On a scale drawing the ratio is $1 : n$. How do you find the length on a drawing?	Divide the real-life length by n.
9	What does a ratio of $1 : 100$ mean on a scale drawing?	1 unit in the drawing is equal to 100 units in real life.
10	In a particular city the ratio of red cars to blue cars is $r : b$. What is the ratio of blue cars to red cars?	$b : r$

Previous questions

Now go back and use these questions to check your knowledge of previous topics.

	Questions	Answers
1	How do you divide by a fraction?	Multiply by the reciprocal of the fraction.
2	What is the highest common factor of two or more numbers?	The largest number that is a factor of all the numbers.
3	How do you convert a fraction into a decimal?	Divide the numerator by the denominator.
4	What is an improper fraction?	A fraction where the numerator is bigger than or equal to the denominator.
5	What is the midpoint formula for finding the midpoint of two points (x_1, y_1) and (x_2, y_2)?	$\text{midpoint} = \left(\frac{x_1 + x_2}{2}, \frac{y_1 + y_2}{2}\right)$

Practice

Exam-style questions

16.1 The lengths of the sides of a right-angled triangle are in the ratio 5 : 12 : 13. The length of the shortest side is 2 cm.

Work out the perimeter of the triangle. **[3 marks]**

16.2 The ratio of pencils to erasers is 7 : 2. There are 90 more pencils than erasers.

How many pencils are there? **[2 marks]**

16.3 The points A, B, C and D form a straight line $ABCD$. The ratio of the length of AB to BC is 6 : 5. The ratio of the length of BC to CD is 10 : 13. The total length of AD is 105 cm.

Work out the length of BC. **[3 marks]**

16.4 Kayleigh is two years younger than Hayley. Bailey is twice as old as Kayleigh. The sum of their three ages is 38.

Find the ratio of Bailey's age to Hayley's age to Kayleigh's age. **[4 marks]**

16.5 Carl and Frida are playing a game with 44 cards. Carl has 20% more cards than Frida.

Work out the number of cards they each have. **[3 marks]**

16.6 The ratio of blue cars to red cars in a traffic survey is 2 : 1. In a follow-up survey, the number of blue cars has increased by 50% and the number of red cars has increased by 25%.

What is the new ratio of blue cars to red cars? Write your answer in the form $a : b$, where a and b are integers. **[3 marks]**

16.7 The ratio of integer x to integer y is 3 : 2.
When x is increased by 8 and y is halved, the ratio becomes 4 : 1.

Work out the values of x and y. **[5 marks]**

Exam-style questions

16.8 There are some copper coins and silver coins on a table. Each coin is either large or small. The ratio of copper coins to silver coins is 4 : 3. The ratio of small copper coins to large copper coins is 3 : 7. The ratio of small silver coins to large silver coins is 1 : 2.

Work out what fraction of all of the coins are small coins. **[5 marks]**

16.9 Given that $(2x - 5) : 6 = 1 : (6 - x)$, find the possible values of x. **[4 marks]**

> **EXAM TIP**
> $a : b = c : d$ then $\frac{a}{b} = \frac{c}{d}$

16.10 A world cruise costs £15 285 plus VAT at 20%. Judith pays a one-off deposit followed by 10 equal payments of £1384.20.

Find the ratio of the deposit Judith pays to the total of the 10 equal payments. Give your answer in its simplest form. **[5 marks]**

16.11 Light travels 4.8555×10^9 km in 4.5 hours.
Sound travels 3.7044×10^4 kilometres in 3 hours.

Work out the ratio of the speed of light to the speed of sound. Write your answer in the form $n : 1$, where n is correct to 3 significant figures. **[4 marks]**

16.12 £425 is shared between Pranay, Quinten, Raha and Stephanie. The ratio of the amount Pranay gets to the amount Raha gets is 4 : 5. Quinten gets 1.25 times the amount Pranay gets. Stephanie gets 2.2 times the amount Quinten gets.

Work out the amount of money that Stephanie gets. **[4 marks]**

16.13 A delivery consists of 560 pieces of fruit. Each piece of fruit is either a banana, an apple, an orange or a pear. $\frac{3}{8}$ of the fruit are oranges. 15% of the fruit are bananas. The ratio of the number of apples to the number of pears is 8 : 11.

Work out how many pears there are in the delivery. **[5 marks]**

16.14 Given that $3x^2 : (5x + 4) = 2 : 1$, find the possible values of x. **[4 marks]**

16.15 Deshawn, Amara and Harper share some marshmallows between them in the ratio 3 : 2 : 5. Harper gives five of her marshmallows to Amara, one of which she eats. The ratio of marshmallows is then 4 : 4 : 5.

Work out how many marshmallows each person had to begin with. **[3 marks]**

Questions referring to previous content

16.16 Work out $3\frac{2}{3} \div 2\frac{1}{2}$. Give your answer as a mixed number. **[3 marks]**

16.17 Simplify fully: $\dfrac{x^2 + 7x + 10}{x^2 + 2x - 15} \times \dfrac{x^2 + x - 12}{x^2 + 2x}$. **[4 marks]**

Knowledge

17 Compound measures and multiplicative reasoning

Measures

Convert between metric units by multiplying or dividing by powers of 10.

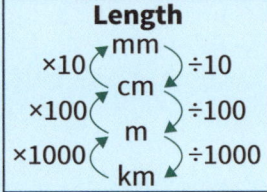
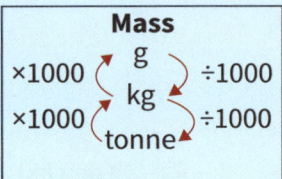
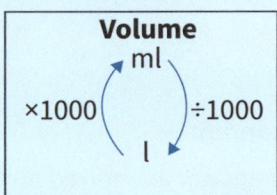

milli- = one thousandth
centi- = one hundredth
kilo- = one thousand

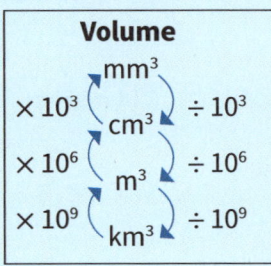
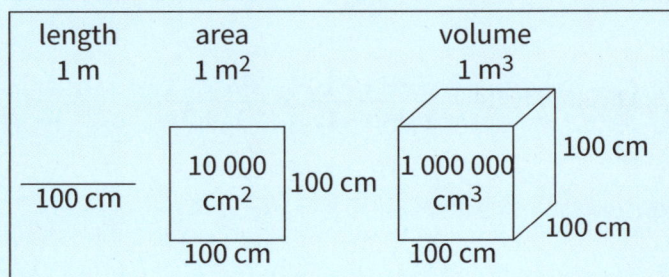

Worked example

A sphere has a volume of 20 cm³ and a surface area of 36 cm².

1. Write the surface area in mm²

    Surface area = $36 \times 10^2 = 3600$ mm² ← To convert from cm² to mm², multiply by 10^2

2. Write the volume in mm³.

    Volume = $20 \times 10^3 = 20\,000$ mm³ ← To convert from cm³ to mm³, multiply by 10^3

Compound measures

A **compound measure** links two measurements. For example, a rate of pay could be £11 per hour and a rate of flow of water could be 3 litres per minute.

You need to know the formulae for these compound measures:

$$\text{speed} = \frac{\text{distance}}{\text{time}} \qquad \text{density} = \frac{\text{mass}}{\text{volume}} \qquad \text{pressure} = \frac{\text{force}}{\text{area}}$$

Worked example

Imani is buying a car. She compares the maximum speeds of two cars.

car 1: top speed = 195 km/h
car 2: top speed = 52 m/s

Which car has the highest top speed?

To compare top speeds, we need to convert the speeds to the same units. Here we will convert car 1 to m/s, but you could convert car 2 to km/h.

$195 \times 1000 = 195\,000$ m/h ← × 1000 to convert to m/h.

$\frac{195\,000}{60 \times 60} = 54.17$ m/s ← Divide by 60 to convert from m/h to m/min, and divide by 60 again to convert to m/s.

So car 1 has the higher top speed.

17

Compound measures

Worked example

The density of a piece of wood is 2.4 g/cm³ and its mass is 1.8 kg. Calculate the volume of the wood.

density = mass / volume — Write the formula linking density, mass and volume.

volume = mass / density — Rearrange to make volume the subject.

1.8 kg = 1800 g — Convert kg to g.

volume = 1800 / 2.4

= 750 cm³ — Substitute to find volume.

Direct proportion

When two values are in **direct proportion**, they will increase and decrease at the same rate.

For example, the number of apples and the cost of the apples are in direct proportion. If one apple costs 30p, then:

apples	cost
1	30p
2	60p
7	210p

×7 (×2), ×2, ×7

To solve direct proportion problems, you can use a common factor, or the **unitary method**:

- Divide to calculate the value of 1 unit.
- Multiply, if necessary, to find the value required.

The symbol for 'proportional to' is ∝.

Worked example: Common factors

1. A pack of 20 biscuits costs £1.50 and a pack of 35 of the same biscuits costs £2.80. Which is the better value?

small pack		large pack	
20	£1.50	35	£2.80
5	£0.375	5	£0.40

   ÷4, ÷4, ÷7, ÷7

   5 is a common factor.

   The smaller pack is better value.

2. A recipe for 12 portions of cheesecake requires 30 biscuits. How many biscuits are required for 32 portions of cheesecake?

portions	biscuits
12	30
4	10
32	80

   ÷3, ×8, ÷3, ×8

   80 biscuits are required for 32 portions.

Worked example: Unitary method

A pack of 240 teabags costs £2.25 and a pack of 150 of the same teabags costs £1.10. Which is the better value?

240 pack		150 pack	
bags	price	bags	price
240	£2.25	150	£1.10
1	£0.009375	1	£0.0073

Divide the price by the number of bags to get the price per unit.

The 150 pack is better value. — Compare the price per unit; lowest is best value.

Knowledge

17 Compound measures and multiplicative reasoning

Graphs of direct proportion

If x is directly proportional to y, then $y = kx$, where k is the **constant of proportionality**.

A graph showing two variables in direct proportion is a straight line through the origin with gradient k.

> **LINK**
> To remind yourself about straight-line graphs, look back at Chapter 6.

Worked example

The graph shows the exchange rate between pounds (£y) and dollars ($$x$) on a given day.

(a) Find an equation for y in terms of x.

gradient $= \dfrac{8}{10} = 0.8$, so $y = 0.8x$

(b) Use your equation to find how much $15 is in pounds.

$y = 0.8 \times 15 = 12$ so £12.

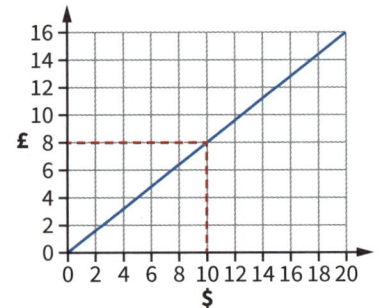

Equations for direct proportion

y can also be directly proportional to a function of x, for example, $y = kx^2$ or $y = k\sqrt{x}$

Worked example

b is proportional to a^2

$b = 20$, when $a = n$

Show that when $a = 3n$, $b = 180$

$b \propto a^2$ so $b = ka^2$ ← Write an equation with k as the constant of proportionality.

$20 = kn^2$ ← Substitute $b = 20$, $a = n$

$k = \dfrac{20}{n^2}$ ← Rearrange to give an expression for k.

$b = \left(\dfrac{20}{n^2}\right)a^2$

$= \dfrac{20a^2}{n^2}$ ← Substitute $k = \dfrac{20}{n^2}$ into $b = ka^2$

$= \dfrac{20(3n)^2}{n^2}$ ← Substitute $a = 3n$ into $b = \dfrac{20a^2}{n^2}$

$= \dfrac{180(n)^2}{n^2}$

$= 180$

17

Inverse proportion

Two values are in **inverse proportion** if, when one *increases*, the other *decreases* at the same rate.

For example, the number of taps and time taken to fill a sink are in inverse proportion.

time	taps
6 mins	1
3 mins	2
1.5 mins	4

÷2, ÷2 on time; ×2, ×2 on taps

Notice that the product of the two values is constant.

Worked example

It takes 8 cleaners 3 hours to clean a block of offices.

1. How long does it take 6 cleaners to clean the offices?

cleaners	hours	cleaner-hours
8	3	8 × 3 = 24
6	$\frac{24}{6} = 4$	24

— Work out how many cleaner-hours the offices require.

— Work out how many hours are needed if there are 6 cleaners.

It takes 6 cleaners 4 hours to clean the offices.

2. How many cleaners are needed to clean the offices in 1.5 hours?

cleaners	hours	cleaner-hours
8	3	24
16	1.5	24

×2 on cleaners; ÷2 on hours

— Work out how many cleaners are needed if there are 1.5 hours. (16 × 1.5 = 24)

16 cleaners are needed to clean the offices in 1.5 hours.

Graphs of inverse proportion

A graph showing inverse proportion will lie close to the x and y axes, and curve away from the origin.

If x is inversely proportional to y, then $y = \frac{k}{x}$ where k is any number and is called the **constant of proportionality**.

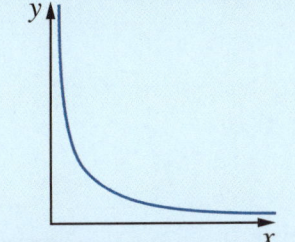

LINK

To remind yourself about reciprocal graphs, look back at Chapter 13.

Equations for inverse proportion

It is also possible for y to be inversely proportional to a function of x.
For example,

$y = \frac{k}{\sqrt{x}}$, here y is inversely proportional to the square root of x.

Worked example

y is inversely proportional to x^2

$y = 2$ when $x = 3$

Find the value of y when $x = 4$

$y = \frac{k}{x^2}$

$2 = \frac{k}{9}$

$k = 18$

When $x = 4$,

$y = \frac{18}{4^2}$

$= \frac{9}{8}$

— Write an equation connecting x and y.

— Substitute x and y and rearrange to find k.

— Substitute k and x into the equation and solve for y. Remember to simplify the fraction.

17 Knowledge 129

Knowledge

17 Compound measures and multiplicative reasoning

Problem solving with proportion

Worked example

m is directly proportional to v^3

When $v = 2$, $m = 24$

v is inversely proportional to n^2

When $n = 10$, $v = 4$

Find a formula for m in terms of n.

$m \propto v^3$ $v \propto \dfrac{1}{n^2}$

$m = kv^3$ — Write the equations with a constant of proportionality, k. $v = \dfrac{k}{n^2}$

$24 = k \times 2^3$ — Substitute given values. $4 = \dfrac{k}{10^2}$

$k = \dfrac{24}{2^3} = 3$ — Solve for k. $k = 4 \times 10^2 = 400$

So $m = 3v^3$ So $v = \dfrac{400}{n^2}$

$m = 3\left(\dfrac{400}{n^2}\right)^3$ — Substitute $v = \dfrac{400}{n^2}$ into $m = 3v^3$.

$= \dfrac{192\,000\,000}{n^6}$

Key terms — Make sure you can write a definition for these key terms: compound measure, constant of proportionality, direct proportion, inverse proportion, unitary method

Retrieval — 17

17 Compound measures and multiplicative reasoning

Learn the answers to the questions below, then cover the answers column with a piece of paper and write as many as you can. Check and repeat.

Questions / Answers

#	Question	Answer
1	How do you convert mm into m?	Divide by 1000 (divide by 10 to convert to cm and then by 100 to convert to m).
2	How do you convert from kg to g?	Multiply by 1000.
3	How do you convert from ml to l?	Divide by 1000.
4	How do you convert cm² into mm²?	Multiply by $10^2 = 100$.
5	How do you convert km² into cm²?	Multiply by $100\,000^2 = 10\,000\,000\,000$ (Multiply by $1000^2 = 1\,000\,000$, and then by $100^2 = 10\,000$).
6	What is the formula for speed?	speed = distance ÷ time
7	What is the formula for density?	density = mass ÷ volume
8	What is the formula for pressure?	pressure = force ÷ area
9	How can you calculate time if you know the distance and speed?	time = distance ÷ speed
10	How do you convert km/h to m/s?	Multiply by 1000 and then divide by 60^2.
11	What would a graph showing direct proportion look like?	A straight line through the origin.
12	What is the equation for direct proportion?	$y = kx$
13	What is the equation for inverse proportion?	$y = \dfrac{k}{x}$

Previous questions

Now go back and use these questions to check your knowledge of previous topics.

#	Question	Answer
1	What does it mean to factorise fully?	Make sure you have used the highest common factor, not just a common factor.
2	What is a variable?	A letter whose value can change.
3	What is an identity?	An equation that is true for all values.

Practice

Exam-style questions

17.1 Water is poured into an empty tank at a rate of 20 cm³/s.

Calculate the time it will take for the tank to contain 2400 cm³ of water in seconds. **[2 marks]**

17.2 5 m³ of wrought iron has mass 38 700 000 g.

Calculate the density of the wrought iron in kg/m³. **[3 marks]**

17.3 (a) A garden snail travels 5.64 m at a speed of 0.047 km/h.

Work out how long it takes. Give your answer in minutes and seconds. **[4 marks]**

> **EXAM TIP**
> Think about units.

(b) A slug travels 0.78 m per minute.

Determine whether the garden snail or the slug is faster. Justify your answer. **[3 marks]**

17.4 'Dogs Love Bach' sell dog food in 3 kg bags. They are offering three bags for £12.99. 'Woof & Ready' sell the same dog food in 2 kg bags. They are offering four bags for £11.00.

Which dog food seller is offering better value? Show your working. **[3 marks]**

17.5 A 12 acre plot of land in France costs 58 800 euros. A 15 acre plot of land in Argentina costs 4 520 000 Argentine pesos. 1 euro = 64.19 Argentine pesos.

Work out which plot of land costs less per acre. **[3 marks]**

17.6 A team of 12 painters can paint a house in 6 days.

(a) Calculate how long it would take 18 painters. **[2 marks]**

(b) Work out how many painters would be needed to paint the house in 3 days. **[2 marks]**

(c) Sketch a graph to show the relationship between the number of painters and the time it takes to paint the house. **[2 marks]**

17.7 T is inversely proportional to W. $W = 4$ when $T = 5$.

Find the value of T when $W = 8$. **[2 marks]**

17.8 p is directly proportional to q. When $q = 4$, $p = 50$.

(a) Find a formula for p in terms of q. **[2 marks]**

(b) Find the value of q when $p = 40$. **[2 marks]**

17.9 f is inversely proportional to g. g is directly proportional to h^2.

Given that $g = 3$ and $h = 0.5$, when $f = 6$, find a formula for f in terms of h. [4 marks]

17.10 A team of 7 workers can complete a building job in 2 days.

(a) How long will it take to complete the job if only 4 of the workers are available? [2 marks]

(b) State one assumption that you made in working out your answer to part (a). [1 mark]

17.11 An object with volume 100 cm³ has mass of 30 kg. Pat says, 'If the volume was increased by 40 cm³ and the mass was increased by 40 kg, the density would increase by more than 60%.'

Is Pat correct? You must show working to support your answer. [2 marks]

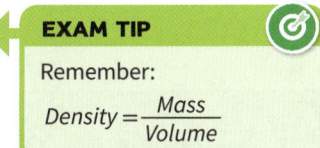

EXAM TIP

Remember:

$$\text{Density} = \frac{\text{Mass}}{\text{Volume}}$$

17.12 f is directly proportional to $\sqrt{g}$. When f is 2, $g = 324$.

g is inversely proportional to h^2. When g is 225, $h = 0.2$.

Find a formula for f in terms of h. Give your answer in its simplest form. [5 marks]

17.13 v is inversely proportional to w^2. When $v = 2$, $w = 3x$.

Show that $v = 0.72$ when $w = 5x$. [3 marks]

17.14 A cheetah can travel 1.98 km in 1 minute. The distance is measured correctly to the nearest 10 m. The time is measured to the nearest 5 seconds.

By considering bounds, work out the average speed in m/s of the cheetah to a suitable degree of accuracy. You must show all your working and give a reason for your answer. [5 marks]

Questions referring to previous content

17.15 The mass of an egg decreases by 11% when you remove the shell. An egg without a shell has a mass of 44.5 g.

Calculate the mass of the egg with the shell on. [3 marks]

17.16 Fred and Ted have some money in the ratio 2 : 1

They both pay £9 for lunch. The ratio of their money is now 5 : 2

How much money did they each have originally? [4 marks]

Knowledge

18 Polygons, angles, and parallel lines

Types of angle

 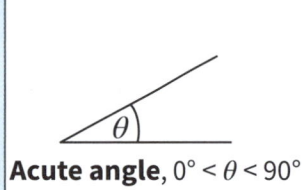 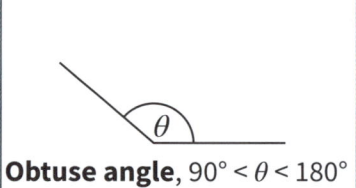 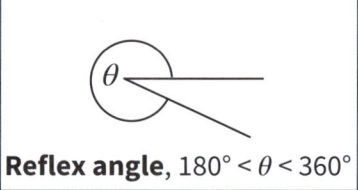

| Right angle, $\theta = 90°$ | Acute angle, $0° < \theta < 90°$ | Obtuse angle, $90° < \theta < 180°$ | Reflex angle, $180° < \theta < 360°$ |

Two lines are **perpendicular** when the angle between them is exactly 90°

Angles can be measured in a clockwise ↻ or anticlockwise ↺ direction.

Angle facts

Vertically **opposite angles** are equal.

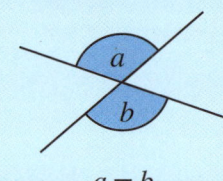

$a = b$

Parallel lines are the same distance from each other. Parallel lines never meet. They are shown on diagrams by arrows on the lines.

Angles a and b are **corresponding angles**. Corresponding angles are always equal.

Angles c and d are **alternate angles**. Alternate angles are always equal.

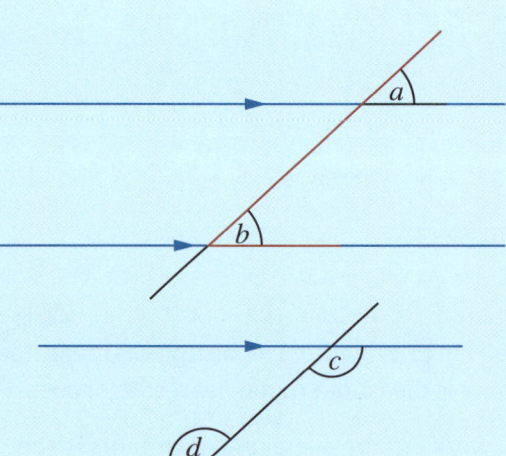

> **WATCH OUT**
>
> In your exams, make sure you use the correct description (alternate angles, corresponding angles, angles at a point, and so on) or you could miss out on marks.

Worked example

Find the values of x, y and z in the diagram.
Give reasons for your answers.

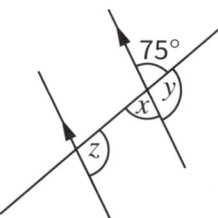

$x = 75°$ — Vertically opposite angles are equal.

$y = 180° - 75°$ — Angles on a straight line add up to 180°.
$ = 105°$

$z = y$ — Corresponding angles are equal.
$ = 105°$

18

Angles in quadrilaterals

A **quadrilateral** is a 2D shape that has four straight sides.

Angles in a quadrilateral add up to 360°

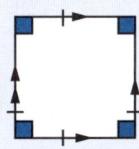 Square   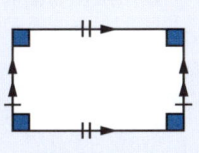 Rectangle   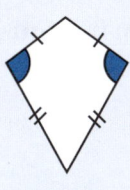 Kite   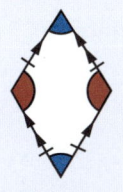 Rhombus   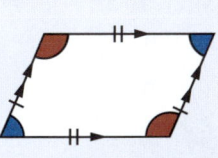 Parallelogram   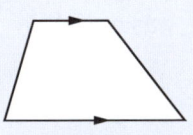 Trapezium

Worked example

Shape $ABCD$ is a parallelogram.
Work out the values of x and y.

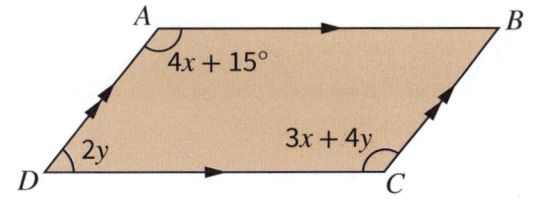

$\angle DAB = \angle BCD$ — Opposite angles in a parallelogram are equal.

$4x + 15 = 3x + 4y$

$4y - x = 15$ ① — Label this equation ①.

$\angle ABC = 2y$ — Opposite angles in a parallelogram are equal.

So:

$(4x + 15) + 2y + (3x + 4y) + 2y = 360$ — Angles in a parallelogram add to 360°.

$7x + 8y + 15 = 360$

$7x + 8y = 345$ ② — Label this equation ②.

$4y - x = 15$ — Write ① and ② as simultaneous equations.
$7x + 8y = 345$

$4y - x = 15$
$x = 4y - 15$ ③ — Rearrange ① to make x the subject.

$7x + 8y = 345$
$7(4y - 15) + 8y = 345$ — Substitute $x = 4y - 15$ into ②, and solve to find the value of y.
$28y - 105 + 8y = 345$
$36y = 450$
$y = 12.5$

$y = 12.5$
$x = 4 \times 12.5 - 15$ — Substitute the value of y into ③ to find the value of x.
$x = 35$

18 Knowledge

Knowledge

18 Polygons, angles, and parallel lines

Angles in polygons

A polygon is a two-dimensional shape with three or more straight edges.

A **regular polygon** (such as an equilateral triangle) has equal sides and equal angles.

An **irregular polygon** has at least one side or angle that is different from the others.

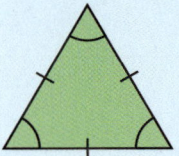 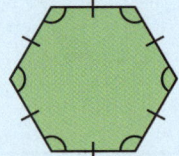 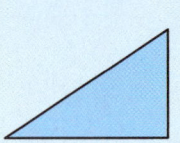 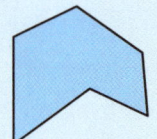

Interior and exterior angles

An **interior angle** is the angle inside a polygon at a **vertex**.

An **exterior angle** is the angle between a side of the shape and a line extended from the side.

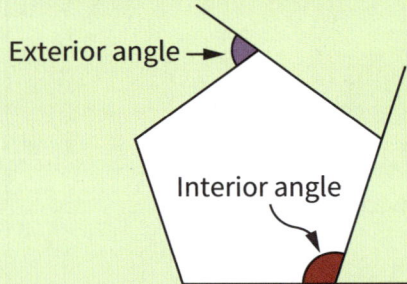

sum of interior angles = (number of sides − 2) × 180°

The number of triangles you can split a shape into is always two fewer than the number of sides.

The angles in a triangle add up to 180°.

There are six triangles in an octagon, so

sum of interior angles = 6 × 180°
$\phantom{\text{sum of interior angles}}$ = 1080°

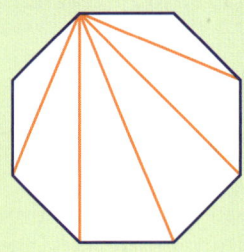

Worked example

A heptagon has interior angles as shown. Calculate the size of the angle marked x.

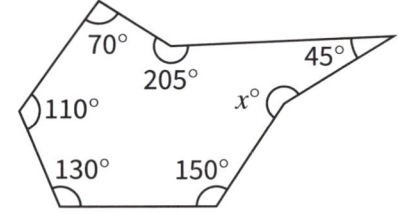

sum of interior angles
$\phantom{\text{sum of interior}}= (7 - 2) \times 180°$
$\phantom{\text{sum of interior}}= 5 \times 180°$
$\phantom{\text{sum of interior}}= 900°$

$70° + 205° + 45° + 150° + 130° + 110° = 710°$

$x = 900° - 710°$
$ = 190°$

Add the six angles you know.

Subtract the sum of the known angles from the sum of all angles.

18

Interior and exterior angles

The sum of the exterior angles is the same for any polygon.

sum of exterior angles = 360°

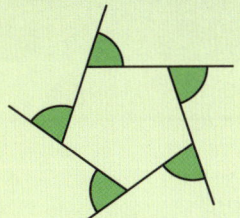

> **REVISION TIP**
>
> The number of exterior angles in a polygon is the same as the number of sides.
>
> The sum of exterior angles = 360°, so
>
> In a regular polygon, all exterior angles are equal so
>
> number of sides × exterior angle = 360°

Worked example

The exterior angle of a regular polygon is 15°.

How many sides does the polygon have?

number of sides × exterior angle = 360°

number of sides = $\dfrac{360°}{\text{exterior angle}}$ ← Rearrange to make 'number of sides' the subject.

number of sides = $\dfrac{360°}{15}$ ← Substitute

= 24

Worked example

The diagram shows a regular hexagon.
Find the size of angle x.

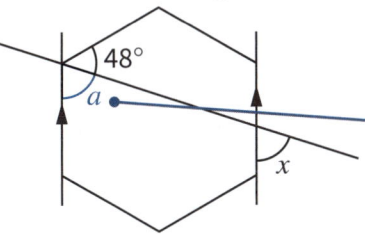

← Find the angle that corresponds to x and draw it on the diagram.

interior angle of a hexagon = $a + 48°$ ← Write the values you know in an equation.

sum of interior angles = $(6 - 2) \times 180°$

= 720° ← Use the sum of interior angles to find the value of one interior angle.

So, one interior angle = $\dfrac{720°}{6} = 120°$

$a + 48° = 120°$

$a = 120° - 48°$ ← Substitute to find value of a.

= 72°

Corresponding angles are equal, so:

$x = a$ ← Find x.

= 72°

18 Knowledge

Knowledge

18 Polygons, angles, and parallel lines

Interior and exterior angles

You can also use the exterior angle to find the interior angle.

Worked example

Find the interior angle of a regular heptagon.

exterior angle = $\dfrac{360°}{\text{number of sides}}$

$a = \dfrac{360°}{7}$ — A heptagon has seven sides.

$= 51.4°$

$a + b = 180°$ — a and b lie on a straight line.

$b = 180° - 51.4°$

$= 128.6°$ — Substitute $a = 51.4°$ to find b.

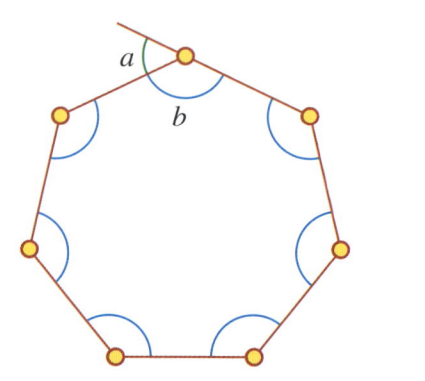

Worked example

The diagram shows a regular octagon.

Calculate the value of x.

Give a reason for each stage of your working.

sum of interior angles = (number of sides − 2) × 180°

Octagon has 8 sides, so

sum of interior angles = (8 − 2) × 180°

$= 1080°$

The octagon is regular, so each interior angle is the same. So,

$\dfrac{1080°}{8} = 135°$

$\angle CBA = 135°$

CB and BA are both sides of the regular octagon, so $CB = BA$.

So CBA is an isosceles triangle.

So $\angle BCA = \angle BAC = x$

$x + x + 135° = 180°$ — Angles in a triangle sum to 180°

$2x = 45°$

$x = 22.5°$

Key terms — Make sure you can write a definition for these key terms

acute angle   alternate angle   corresponding angle
exterior angle   interior angle   irregular polygon
obtuse angle   parallel   quadrilateral   regular polygon
perpendicular   reflex angle   right angle   transversal   vertex

Retrieval 18

18 Polygons, angles, and parallel lines

Learn the answers to the questions below, then cover the answers column with a piece of paper and write as many as you can. Check and repeat.

Questions | Answers

#	Question	Answer
1	What is an acute angle?	An angle that is less than 90°.
2	What is a reflex angle?	An angle that is greater than 180° and less than 360°.
3	What is an interior angle?	The angle between two lines on the inside of a shape.
4	What do angles at a point add up to?	360°
5	What do angles on a straight line add up to?	180°
6	What do the angles in a triangle sum to?	180°
7	What are the properties of a trapezium?	A trapezium is a quadrilateral with one pair of parallel sides.
8	What is a polygon?	A 2D shape with three or more straight sides.
9	What is a regular polygon?	A polygon with sides of equal length and angles of equal size.
10	What is the formula for the sum of interior angles in a polygon?	sum of interior angles = (number of sides − 2) × 180°
11	What do the exterior angles of any polygon add up to?	360°
12	If you are given one exterior angle of a regular polygon, how do you find the number of sides it has?	360 ÷ size of the exterior angle

Previous questions

Now go back and use these questions to check your knowledge of previous topics.

Questions | Answers

#	Question	Answer
1	What is a mixed number?	A number and a fraction together.
2	What percentage is equivalent to $\frac{1}{4}$?	25%
3	Why is 2 the only even prime number?	All other even numbers are divisible by 2 and therefore have more than two factors.
4	What does the gradient of the line measure?	How steep it is.
5	How do you find the next term in a geometric sequence?	You multiply by the common ratio.

Practice

Exam-style questions

18.1 ABCG is a parallelogram.

Work out the size of angle x in the diagram. Give a reason for each stage of your working. [3 marks]

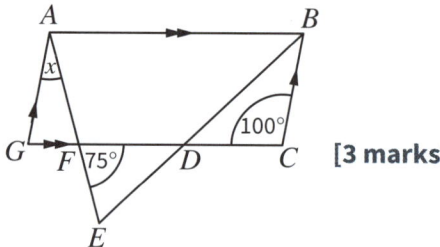

18.2 The angles in a parallelogram are in the ratio $1:3:1:x$.

(a) Write down the value of x. [1 mark]

(b) Work out the sizes of all four angles. [2 marks]

EXAM TIP
The sum of the angles in a quadrilateral is 360°.

18.3 Work out the values of c and d. You must show all your working. [4 marks]

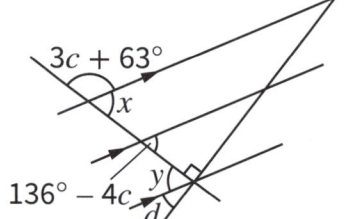

18.4 A circular pie is cut into four pieces, as shown. AC is the diameter of the pie.

Work out the size of the angle for the largest slice of pie. Give a reason for each stage of your working. [4 marks]

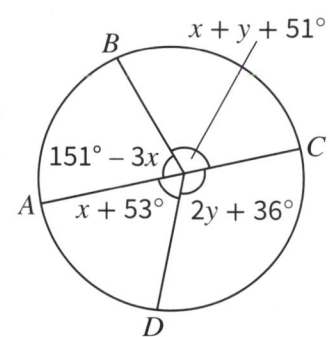

18.5 The quadrilateral ABCD is a rhombus.

Work out the value of angle x, giving full reasons for each step of your working. [3 marks]

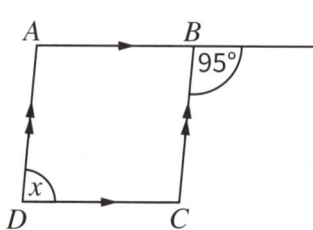

18.6 In the diagram, angle BFA is 44°.

Work out the size of angle FDE. [3 marks]

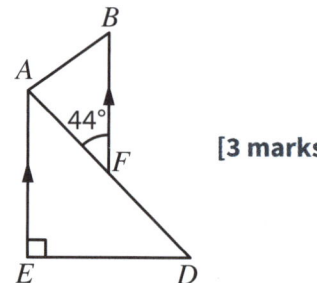

18.7 *PQR* and *STU* are both equilateral triangles of the same size. *T* is the midpoint of *PR*. *Q* is the midpoint of *SU*.

Work out the size of angle *t*. Give reasons for each step of your working. [3 marks]

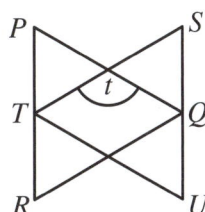

18.8 Work out the values of *x* and *y* for the kite shown. [4 marks]

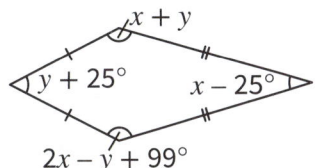

18.9 Part of a regular polygon is shown, along with one exterior angle.

Write the name of the polygon. Give a reason for your answer. [2 marks]

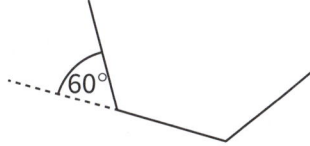

18.10 The sum of the interior angles of a polygon is 1620°.

Work out how many sides the polygon has. [2 marks]

18.11 The diagram shows a square, two equilateral triangles and two regular octagons.

Work out the size of angle *x*. [4 marks]

EXAM TIP
Consider the rule for the sum of angles round a point.

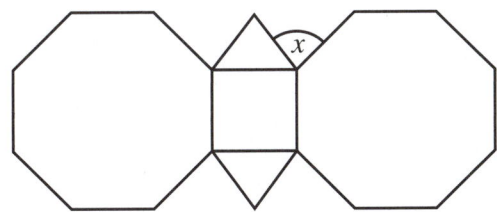

18.12 Jemima says she has drawn a regular polygon with an interior angle of 80°. Sophia says that it is impossible.

Who is correct? Justify your answer. [3 marks]

18 Practice 141

Exam-style questions

   18.13 *ABC* and *DEFG* are parallel lines. Work out the size of angle *x*. Give a reason for each stage of your working.

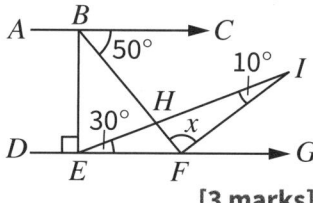

[3 marks]

18.14 *ABCDEF* is a hexagon. Angle *BCD* = 2 × Angle *FED*.
Work out the size of angle *FED*.
Show your working.

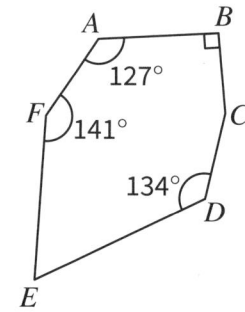

[5 marks]

18.15 *ABCDE* is a pentagon. The pentagon has one line of symmetry.
AB = *AE* and *BC* = *ED*.
Angle *ABC* = 1.5 × Angle *BAE*.

Work out the size of angle BCD. Show your working.

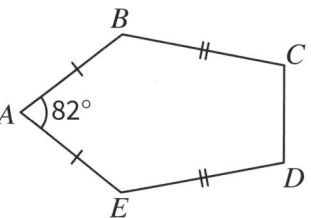

Questions referring to previous content

   18.16 Solve $4^2 \times 8^2 = \dfrac{1}{2^x}$ [4 marks]

   18.17 The variable *y* is indirectly proportional to $\sqrt{2x}$

$y = \dfrac{1}{4}$ when *x* = 32. Find the value of *x* that gives $y = 3\sqrt{2}$ [5 marks]

Knowledge 19

19 Area and perimeter (including circles)

Area

The **area** is the size of the space inside a two-dimensional shape.

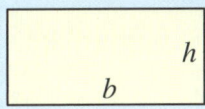
area of rectangle = base × perpendicular height
$= bh$

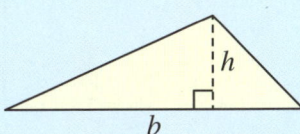
area of triangle $= \dfrac{(\text{base} \times \text{height})}{2}$
$= \dfrac{1}{2}bh$

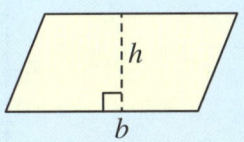
area of parallelogram = base × perpendicular height
$= bh$

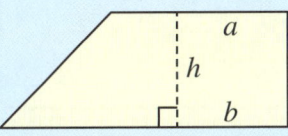
area of trapezium $= \dfrac{1}{2}(a + b)h$

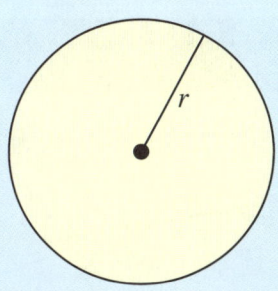
area of circle = π × radius squared
$= \pi r^2$

To find the area of a compound shape:

- Split into simpler shapes.
- Find the area of each shape.
- Add up the separate areas.

Worked example

Calculate the area of this shape.

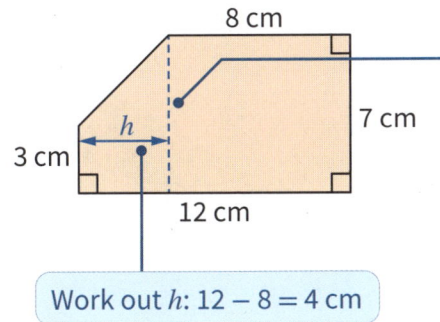

Work out h: $12 - 8 = 4$ cm

Split the shape into a rectangle and a trapezium.

total area = area of trapezium + area of rectangle
$= \dfrac{1}{2} \times 4(3 + 7) + (8 \times 7)$
$= 2(3 + 7) + (8 \times 7)$
$= 2(10) + 56$
$= 76$ cm²

REVISION TIP
There will often be more than one way to split the shape up. Can you think of another way to divide this shape to find its area?

For area, we use units squared because we are multiplying two lengths together.

Knowledge

19 Area and perimeter (including circles)

Area

Worked example

The area of the trapezium is 42 cm². Calculate its perimeter.

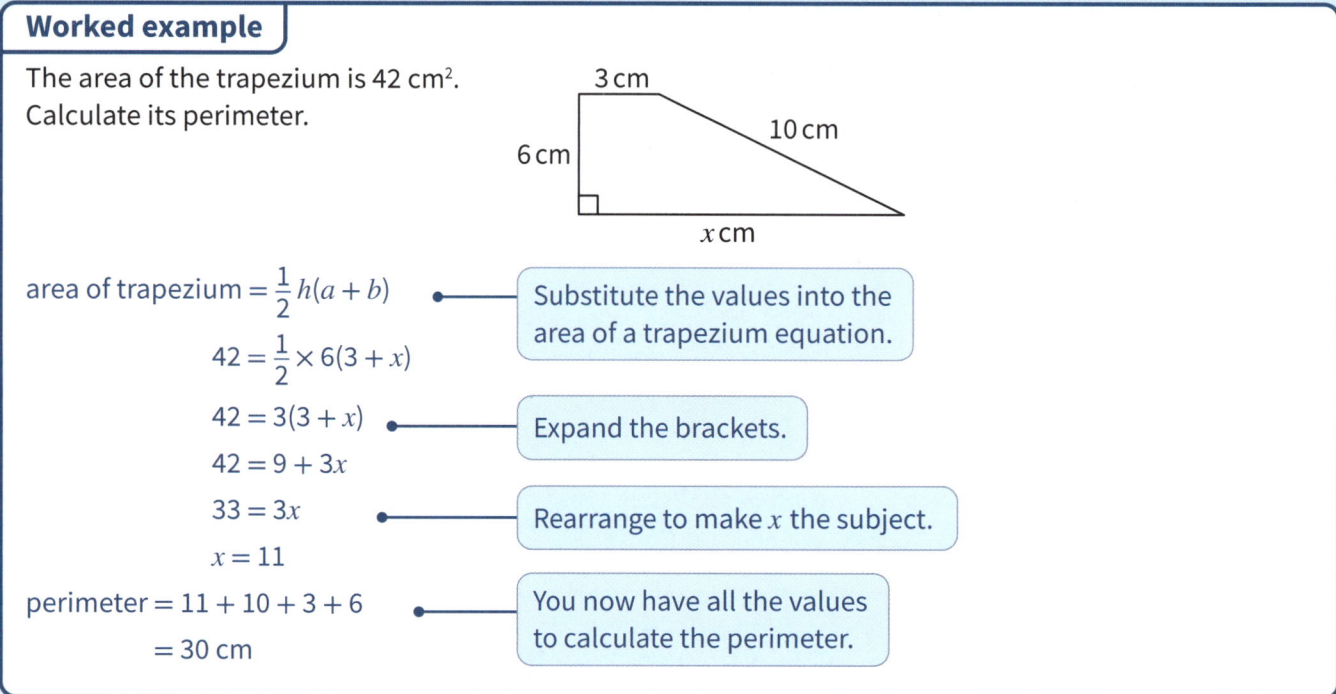

area of trapezium $= \frac{1}{2}h(a+b)$ — Substitute the values into the area of a trapezium equation.

$42 = \frac{1}{2} \times 6(3+x)$

$42 = 3(3+x)$ — Expand the brackets.

$42 = 9 + 3x$

$33 = 3x$ — Rearrange to make x the subject.

$x = 11$

perimeter $= 11 + 10 + 3 + 6$ — You now have all the values to calculate the perimeter.

$= 30$ cm

Parts of a circle

Here are the parts of a circle:

Radius: The distance of a point on the circle to the centre.

Diameter: The distance across the circle through the centre.
diameter = 2 × radius

Circumference: The perimeter of a circle.

Chord: A line that joins two points on a circle.

Segment: The region formed by a chord and the part of the circumference between the two points.

Tangent: A line that touches the circumference of a circle in exactly one point, and is perpendicular to the radius at the point of contact.

Arc: Part of the circumference of the circle.

Sector: Part of the circle formed by two radii and an arc. It is shaped like a slice of cake!

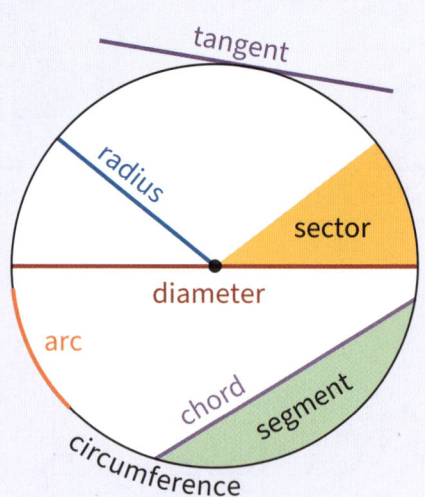

Circle equations

Worked example

A semicircle has radius 4 cm.

(a) Calculate the area of the semicircle. Give your answer correct to 1 d.p.

$$\text{area of circle} = \pi \times 4^2$$
$$= 16\pi \text{ cm}^2$$
$$\text{area of semicircle} = \frac{16\pi}{2}$$
$$= 25.1 \text{ cm}^2 \text{ (to 1 d.p.)}$$

(b) Calculate the perimeter of the semicircle. Give your answer correct to 1 d.p.

$$\text{perimeter of semicircle} = \frac{\text{circumference of circle}}{2} + \text{diameter of circle}$$
$$\text{circumference of circle} = 2 \times \pi \times 4$$
$$= 8\pi \text{ cm}$$
$$\text{circumference of semicircle} = 4\pi$$
$$\text{perimeter of semicircle} = 4\pi + 8$$
$$= 20.6 \text{ cm (to 1 d.p.)}$$

WATCH OUT
Leave your answer in terms of pi until the very end.

- Identify what you need to calculate.
- Calculate the circumference of the circle, and divide it by 2.
- Add the diameter to find the perimeter.

Worked example

Calculate the area of this compound shape, correct to 1 d.p.

$$\text{area of rectangle} = 20 \times 6$$
$$= 120 \text{ cm}^2$$
$$\text{radius of semicircle} = \frac{6}{2}$$
$$= 3 \text{ cm}$$
$$\text{area of semicircle} = \frac{\pi \times 3^2}{2}$$
$$= 14.13 \ldots \text{ cm}^2$$
$$\text{total area} = 120 + 14.134$$
$$= 134.1 \text{ cm}^2 \text{ (to 1 d.p.)}$$

The shape is formed of a rectangle and a semicircle. First calculate the area of the rectangle (base × height).

Now calculate the area of the semicircle.

WATCH OUT
Remember to calculate the radius first. Do not use the diameter.

Knowledge

19 Area and perimeter (including circles)

Arcs and sectors

For a sector with a radius of r and an angle of θ:

area of sector $= \dfrac{\theta}{360} \times$ area of full circle

$= \dfrac{\theta}{360} \times \pi r^2$

arc length $= \dfrac{\theta}{360} \times$ circumference of full circle

$= \dfrac{\theta}{360} \times 2\pi r$

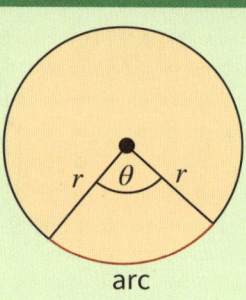

arc

Worked example

1. Calculate the arc length of the sector.

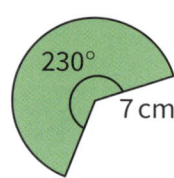

230°

7 cm

arc length $= \dfrac{\theta}{360°} \times$ circumference of full circle — Write out the equation.

arc length $= \dfrac{230°}{360°} \times 2 \times \pi \times 7$ — Substitute values and calculate.

$= 28.1$ cm (to 1 d.p.)

2. Calculate the area of the sector. Give your answers to 1 d.p.

sector area $= \dfrac{\theta}{360°} \times \pi r^2$

$= \dfrac{230°}{360°} \times \pi \times 7^2$

$= 98.3$ cm² (to 1 d.p.)

Worked example

A circle of radius r has a sector of angle 200° removed from it.

The perimeter of the sector is 30 cm.

Calculate the value of r to 1 d.p.

perimeter of sector $=$ arc length $+ 2r$

$30 = \dfrac{200°}{360°} \times 2\pi r + 2r$ — Substitute known values into the equation.

$30 = \dfrac{20}{18} \pi r + 2r$

$30 = \left(\dfrac{20}{18} \pi + 2\right) r$ — Simplify.

$r = \dfrac{30}{\dfrac{20}{18} \pi + 2}$ — Rearrange to make r the subject.

$r = 5.5$ cm (to 1 d.p.)

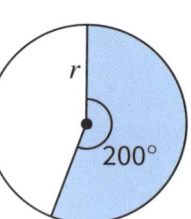

REVISION TIP

A sector can be bigger than a semicircle.

Arcs and sectors

Worked example

OAB is a sector of a circle with centre O and radius 9 m.
The area of the sector is 40.5 m².
Calculate the perimeter of OAB.

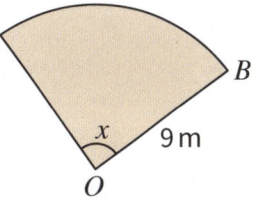

$\dfrac{x}{360°} \times \pi \times 9^2 = 40.5$ — You know the area of the sector, so substitute the values you know into the equation for the area of a sector.

$x = \dfrac{40.5 \times 360°}{\pi \times 9^2}$

$x = 57.3295°$ — Rearrange to make x the subject, and do the calculation to find the value of the angle.

arc length $= \dfrac{57.3295° \ldots}{360°} \times 2\pi \times 9$

$= 9$ m — Calculate the length of the arc.

perimeter of sector $=$ radius $\times 2 +$ length of arc

$= 9 \times 2 + 9$

$= 27$ m

Key terms: Make sure you can write a definition for these key terms

arc   area   chord   circumference   diameter
radius   sector   tangent

19 Knowledge

Retrieval

19 Area and perimeter (including circles)

Learn the answers to the questions below, then cover the answers column with a piece of paper and write as many as you can. Check and repeat.

Questions | Answers

1. What is the formula used to find the area of a trapezium? — $\frac{1}{2} \times (a+b)h$

2. What is the formula used to find the area of a parallelogram? — length of base × perpendicular height

3. What is a compound shape? — A shape made up of two or more simpler shapes.

4. How can you find the area of a compound shape? — Split it into simpler shapes, find the area of each shape and then add the areas together.

5. What is a line that touches a circle in exactly one point called? — A tangent.

6. What is the radius? — The distance from the centre of a circle to the edge.

7. What is a sector? — A 'slice' of a circle between two radii and an arc.

8. How do you find the diameter when you know the radius? — Multiply the radius by 2.

9. What is a chord? — A line that joins two points on a circle.

10. What is the formula used to find the area of a circle? — $A = \pi r^2$

11. What is the formula used to find the circumference of a circle? — $C = \pi d$ or $C = 2\pi r$

12. Describe how you find the area of a semicircle. — Find the area of the circle using $A = \pi r^2$ and divide the answer by 2.

Previous questions

Now go back and use these questions to check your knowledge of previous topics.

Questions | Answers

1. What is rounding? — Writing a number to a given accuracy.

2. What is the top number in a fraction called? — The numerator.

3. What is the general equation of a cubic graph? — $y = ax^3 + bx^2 + cx + d$

4. How do you find the next term of an arithmetic sequence? — Add the common difference.

Exam-style questions

19.1 The perimeter of a right-angled triangle is 90 cm. The lengths of its sides are in the ratio 5 : 12 : 13.

Work out the area of the triangle. [3 marks]

 19.2 A circle has an area of 25π cm².

What is the circumference of the circle in terms of π? [2 marks]

EXAM TIP
First find the radius of the circle.

19.3 Calculate the area and perimeter of this semicircle. Give your answers correct to 1 decimal place. [2 marks]

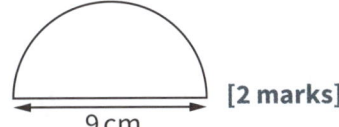

19.4 The circular disc shown has been cut into six identical slices. The radius of the disc is 18 cm.

(a) Write the size of angle x. [1 mark]

(b) Work out the area of a single sector of the disc, giving your answer in terms of π. [2 marks]

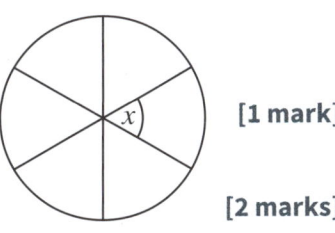

19.5 This sculpture is made using three semicircles and is arranged as shown in the diagram.

Work out the perimeter of the sculpture, giving your answer to 3 significant figures. [3 marks]

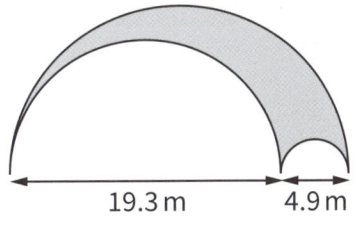

EXAM TIP
You will need to know the diameters of all 3 circles.

19.6 This compound shape is made using a semicircle on top of an isosceles trapezium.

Work out the area of the compound shape to 1 decimal place. [4 marks]

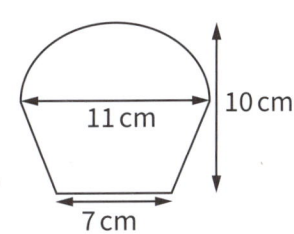

Exam-style questions

19.7 A trapezium and a triangle have the same area. The trapezium has parallel sides of length 9.4 cm and x cm, and a height of 12.8 cm. The triangle has a base of length 12.8 cm and a height of 17.9 cm.

Work out the length x. [3 marks]

19.8 The diagram has been created using two identical quarter circles.

Find the exact area of the shaded region. [3 marks]

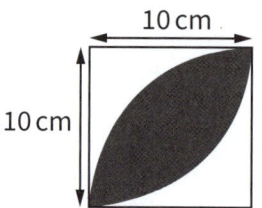

EXAM TIP
You will need to find the area of the quarter circles and the square.

19.9 Giving your answers to 1 decimal place and clearly stating your units, work out

(a) the area of the sector [3 marks]

(b) the perimeter of the sector. [3 marks]

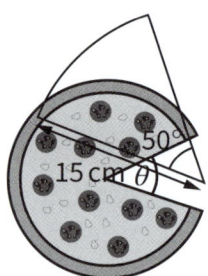

EXAM TIP
The perimeter of a sector is the sum of the arc length and twice the radius.

19.10 A pizza has a radius of 12 cm. Jason takes a slice of the pizza. The remaining pizza has a perimeter of 90 cm.

To the nearest degree, work out the angle, θ, of Jason's slice of pizza. [5 marks]

19.11 AOB is a sector of a circle with radius 17 cm. The area of the sector is 60 cm².

Calculate the perimeter of the sector. Give your answer to 3 significant figures. [4 marks]

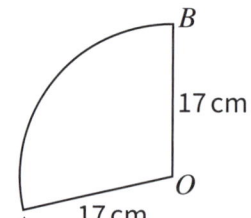

19.12 The diagram shows a rectangle ABCD with length 10 cm. AED and BEC are congruent quarter circles.

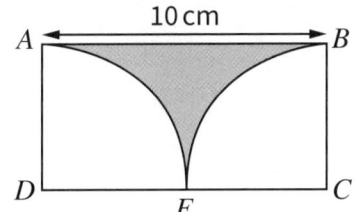

Show that $\dfrac{\text{Area of shaded region}}{\text{Area of rectangle}} = \dfrac{4 - \pi}{4}$. **[4 marks]**

Questions referring to previous content

19.13 Line l_1 intersects the coordinate axes at points A and B as shown. B has coordinates (0, 5). Line l_2 passes through the points $(a, 7)$ and $(7, 16)$ and is perpendicular to line l_1.

The area of triangle OAB is 25 square units.

Work out the value of a. **[4 marks]**

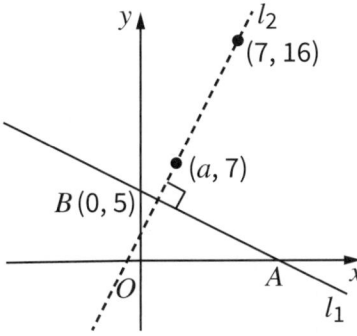

19.14 (a) Write 168 as a product of its prime factors. **[3 marks]**

(b) When a number, n, is cubed, the answer is the same as the answer to 168×441.

Given that $441 = 3^2 \times 7^2$, work out the value of n. **[2 marks]**

Knowledge

20 Surface area and volume

3D shapes

A **three-dimensional (3D) shape** is a solid shape.

3D shapes have **faces**, **edges** and **vertices** (singular: **vertex**).

Faces are the flat surfaces of a 3D shape.

Edges are the lines where two faces meet.

Vertices are the points where two or more edges meet (sometimes called a corner).

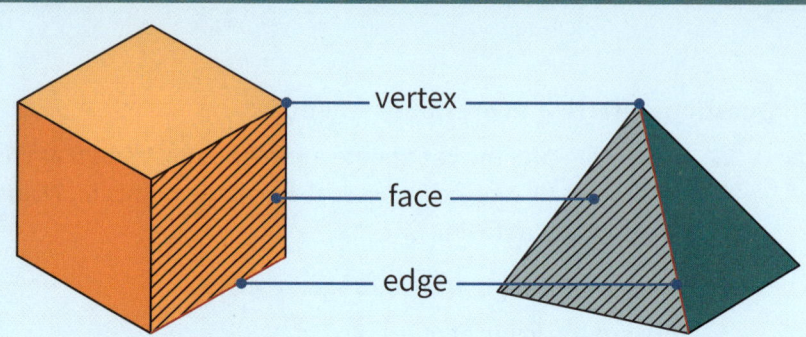

Prisms have a polygon as a base and a constant **cross-section** joined by rectangles. A polygon is any 2D shape with straight sides.

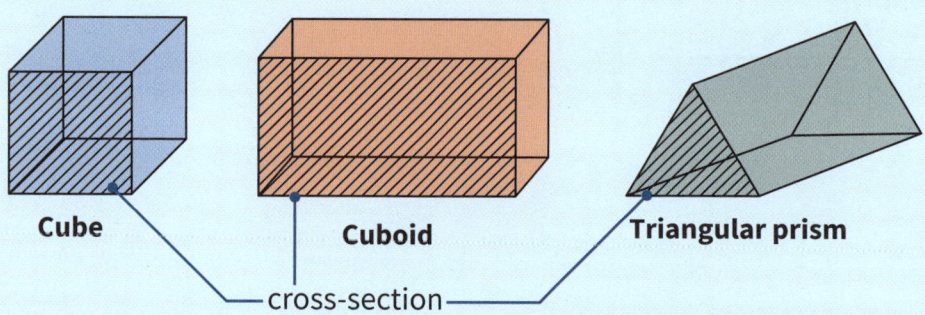

Cube   **Cuboid**   **Triangular prism**

cross-section

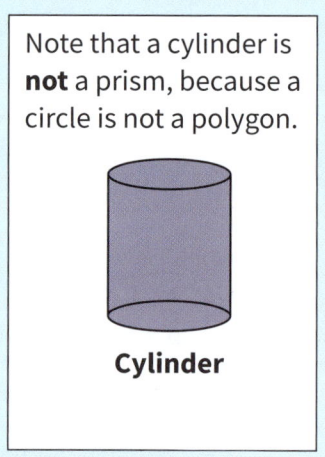

Note that a cylinder is **not** a prism, because a circle is not a polygon.

Cylinder

Note that a constant cross-section means it is the same from one end to the other. Cubes and cuboids have more than one constant cross-section.

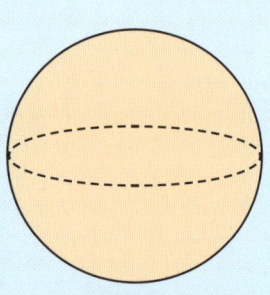

Sphere

A **sphere** is a 3D shape with one curved surface. It is the shape of a ball.

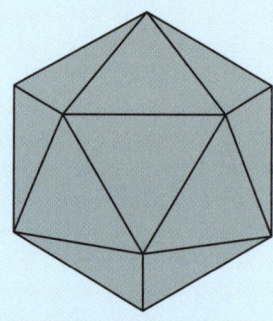

Polyhedron

A **polyhedron** is a solid shape with many faces. This one is called a icosahedron; it has 20 faces.

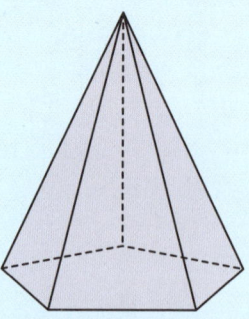

Pyramid

A **pyramid** is a 3D shape with flat faces. It has a polygon base and triangular faces which meet at a point.

20

Volume and surface area formulae

Volume (V) is the amount of space inside a 3D shape. It is measured in cubic units (cm^3, m^3, and so on).

Surface area (SA) is the sum of the area of each face. It is measured in square units (cm^2, m^2, and so on).

Cuboid 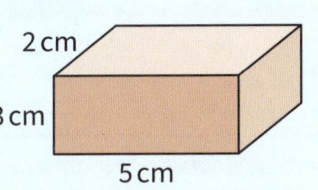	V = length × width × height = 5 × 2 × 3 = 30 cm^3	SA = sum of each face = 2(5 × 2 + 5 × 3 + 2 × 3) = 62 cm^2
Prism 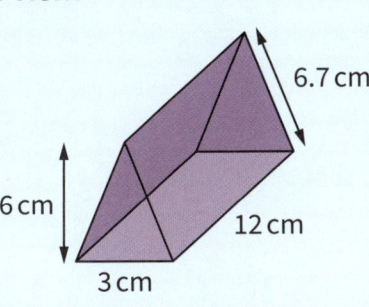	V = area of base × length area of base = $\frac{1}{2}$ × l × w = $\frac{1}{2}$ × 3 × 6 = 9 cm^2 V = 9 × 12 = 108 cm^3	SA = (2 × triangular face) + (2 × sloping face) + base area of one triangular face = $\frac{1}{2}$ × bh = $\frac{1}{2}$ × 3 × 6 = 9 cm^2 area of one sloping face = 12 × 6.7 = 80.4 cm^2 area of base = 12 × 3 = 36 cm^2 SA = (2 × 9) + (2 × 80.4) + 36 = 214.8 cm^2
Cylinder 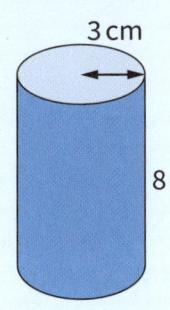	V = area of circular face × height = $\pi r^2 h$ V = $\pi \times 3^2 \times 8$ = 72π cm^3	SA = area of curved surface + 2 × area of circular face area of curved surface = $2\pi rh$ = $2\pi \times 3 \times 8$ = 48π area of circular face = πr^2 = $\pi \times 3^2$ = 9π SA = $48\pi + 2 \times 9\pi$ = 66π cm^2

REVISION TIP

The area of the curved surface is a rectangle.

Its width wraps around the circumference, so its width is $C = 2\pi r$. This is multiplied by the height to get the area of the rectangle.

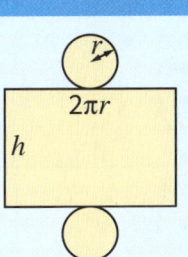

20 Knowledge

Knowledge

20 Surface area and volume

Volume and surface area formulae

Sphere		
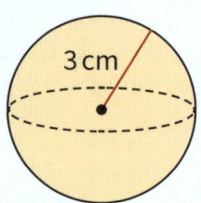 3 cm	$V = \dfrac{4}{3}\pi r^3$ $= \dfrac{4}{3}\pi \times 3^3$ $= 36\pi \text{ cm}^3$	$SA = 4\pi r^2$ $= 4\pi \times 3^2$ $= 36\pi \text{ cm}^2$
Hemisphere		
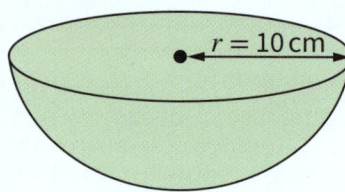 $r = 10$ cm	$V = \dfrac{1}{2} \times$ area of sphere $= \dfrac{1}{2} \times \dfrac{4}{3}\pi r^3$ $= \dfrac{2}{3}\pi r^3$ $= \dfrac{2}{3}\pi \times 10^3$ $= \dfrac{2000\pi}{3} \text{ cm}^3$	SA = area of curved surface + area of base $= \dfrac{1}{2}$ SA of sphere + area of base area of curved surface $= \dfrac{1}{2} \times 4\pi r^2$ area of base $= \pi r^2$ area of curved surface $= \dfrac{1}{2} \times 4\pi r^2$ $= 2\pi \times 10^2$ $= 200\pi \text{ cm}^2$ area of base $= \pi r^2$ $= 100\pi \text{ cm}^2$ SA $= 100\pi + 200\pi$ $= 300\pi \text{ cm}^2$
Cone		
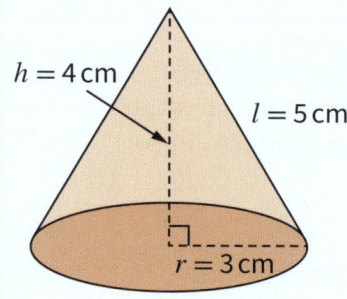 $h = 4$ cm, $l = 5$ cm, $r = 3$ cm	$V = \dfrac{1}{3}\pi r^2 h$ $= \dfrac{1}{3}\pi \times 3^2 \times 4$ $= 12\pi \text{ cm}^3$	SA = area of curved surface + area of base area of curved surface $= \pi r l$ $= \pi \times 3 \times 5$ $= 15\pi \text{ cm}^2$ area of base $= \pi r^2$ $= 9\pi \text{ cm}^2$ SA $= 15\pi + 9\pi$ $= 24\pi \text{ cm}^2$

> **REVISION TIP**
> Make sure you use the vertical height, h, when calculating the volume.

> **REVISION TIP**
> Make sure you use the slant height, l, when calculating the surface area..

Volume and surface area formulae

Pyramid

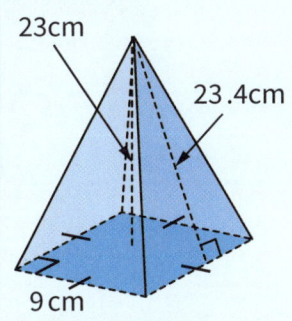

23 cm, 23.4 cm, 9 cm

$V = \frac{1}{3}$ area of base × height

area of base = $l \times w$

$V = \frac{1}{3} (9 \times 9) \times 23$

$= 621 \text{ cm}^3$

SA = area of triangular faces + area of base

area of one triangular face = $\frac{1}{2} \times 9 \times 23.4$

$= 105.3 \text{ cm}^2$

area of base = 9^2

$= 81 \text{ cm}^2$

SA = $4 \times 105.3 + 81$

$= 502 \text{ cm}^2$ (3 s.f.)

Worked example

A sphere has a surface area of $36\pi \text{ cm}^2$. Calculate the volume of the sphere in terms of π.

$4\pi r^2 = 36\pi$ — Substitute $36\pi \text{ cm}^2$ into the formula for the surface area of a sphere.

$\frac{4\pi r^2}{4\pi} = \frac{36\pi}{4\pi}$

$r^2 = \frac{36\pi}{4\pi}$ — Rearrange to make r the subject.

$r^2 = 9$

$r = \sqrt{9}$

$r = 3$

$V = \frac{4}{3} \times \pi \times 3^3$ — Substitute the value of r into the formula for the volume of a sphere

$= 36\pi \text{ cm}^3$ — Leave your answer in terms of π.

Worked example

A fence post is made from a cylinder and a cone.
The height of the cylinder is 10 m.
The radius of the cylinder is 1 m
The height of the cone is 2 m.
Find the volume of the post.

$V_{cylinder}$ = area of circular face × height

$= \pi \times 1^2 \times 10$ — Write the formula for the volume of the cylinder. (Remember: area of a circle is πr^2)

$= 10\pi$ — Substitute the known value and calculate.

$V_{cone} = \frac{1}{3} \times$ area of base × height — Write the formula for the volume of the cone.

$= \frac{1}{3} \times \pi \times 1^2 \times 2$

$= \frac{2}{3}\pi$ — Substitute the known value and calculate.

total volume $= 10\pi + \frac{2}{3}\pi$ — Add the two volumes together.

$= 33.5 \text{ m}^3$

 # Knowledge

20 Surface area and volume

Volume and surface area formulae

Worked example

A cylindrical flask has an internal surface area of 93 000 mm² and a volume of 2.2 litres.

1. What is the volume of the flask in cubic millimetres?
   (2.2 litres = 2200 cm³) — *Convert litres to mm³*
   2200 × 10³ = 2 200 000 mm³

2. The inside of the flask is covered in a special coating which costs 0.3p/cm². Calculate the total cost of the coating.

   $\frac{93\,000 \text{ mm}^2}{10^2} = 930 \text{ cm}^2$ — *Convert the surface area to cm². Multiply by the cost per cm²*

   cost = 930 × 0.3p
   = 279p — *Convert into £*
   = £2.79

REVISION TIP
1 cm³ = 1 ml

Worked example

The diagram shows a metal cuboid.

The cuboid is melted and made into smaller cubes.

Each of the smaller cubes has side length 1.5 cm.

What is the greatest number of cubes that can be made?

Cuboid dimensions: 1 m, 2 m, 3 m

volume of cuboid = 1 × 2 × 3 — *Calculate the volume of the cuboid (length × width × height).*
= 6 m³
= 6 000 000 cm³

volume of one cube = 1.5³ — *Calculate the volume of the smaller cubes (length³).*
= 3.375 cm³

number of cubes = $\frac{\text{volume of cuboid}}{\text{volume of one cube}}$ — *Work out how many cubes can be made.*

= $\frac{6\,000\,000}{3.375}$

= 1 777 777.78

1 777 777 is the greatest number of cubes that can be made.

WATCH OUT
You are looking for the number of *complete* cubes that can be made, so you need to round down not up!

Key terms — Make sure you can write a definition for these key terms.

cross-section   cuboid   cone   cylinder   edge   face   hemisphere
polyhedron   prism   pyramid   sphere   surface area
three-dimensional (3D) shape   vertex   volume

 20 Surface area and volume

Retrieval

20 Surface area and volume

Learn the answers to the questions below, then cover the answers column with a piece of paper and write as many as you can. Check and repeat.

Questions | Answers

1. What is a vertex? — The point where two or more edges meet.
2. What is the same about the cross-section of any prism? — Its cross-section is a polygon.
3. What is a polygon? — Any 2D shape with three or more straight edges.
4. Why is a cylinder not a prism? — Because a circle is not a polygon.
5. What is a face of a 3D shape? — A flat surface.
6. What is an edge of a 3D shape? — Where two faces meet.
7. What is the surface area of a 3D shape? — The sum of the areas of the surfaces of a 3D shape.
8. What is the volume of a 3D shape? — The space inside a 3D shape.
9. How do you find the volume of a cuboid? — Multiply the length by the width by the height.
10. How do you find the volume of a prism? — Multiply the area of the cross-section by the length.
11. What units do you use for volume? — Cubic units.
12. Describe how you find the volume of a cylinder. — Find the area of the circular face then multiply the answer by the height.
13. What is the formula used to find the volume of a pyramid? — volume = $\frac{1}{3}$ × area of base × height
14. How can a net help you to find the surface area of a 3D shape? — You can find the area of each face and add them together.
15. How many faces does a cylinder have? — 2 (the circular ends)

Previous questions

Now go back and use these questions to check your knowledge of previous topics.

Questions | Answers

1. How do you convert from ml to l? — Divide by 1000.
2. What is an acute angle? — An angle that measures more than 0 and less than 90°
3. What does $a^2 - b^2$ factorise to? — $(a-b)(a+b)$
4. To simplify a ratio fully, you divide all the numbers in the ratio by what? — The highest common factor of the numbers.
5. What is the sum of the exterior angles of any polygon? — 360°

Practice

Exam-style questions

20.1 An unopened cylindrical can of baked beans has a height of 11 cm and a radius of 4 cm. Work out

(a) the volume of the can in terms of π [2 marks]

(b) the surface area of the can to 3 significant figures. [4 marks]

EXAM TIP
A closed cylinder has three surfaces.

20.2 The triangular prisms shown have the same volume. Work out the height h. State your units. [4 marks]

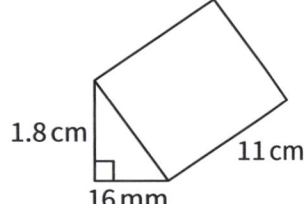
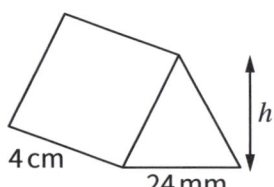

20.3 The volume of a solid wooden cube is $2\sqrt{2}$ m³. A cylindrical hole, of radius 0.25 m, is drilled through the length of the cube and its centre.

Work out the total surface area of the six external faces of the resulting solid. Give your answer in the form $a + b\pi$. [3 marks]

EXAM TIP
Start by finding the length of the cube.

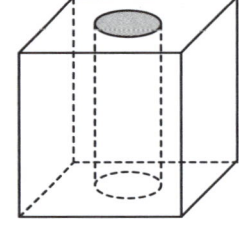

20.4 Karen fills two cylindrical jugs of different sizes with water. The ratio of the volume of the larger jug to the volume of the smaller jug is 2:1.

The total volume of the two jugs is 2400π cm³.

The larger jug has a radius of 12 cm.

(a) Calculate the height of the larger jug to 3 significant figures. [4 marks]

(b) Given that the radius of the smaller jug is equal to its height, calculate the radius of it to 3 significant figures. [3 marks]

158   20 Exam-style questions

20.5 The Great Pyramid of Giza has a square base of length 230 m and a height of 147 m.

Work out the volume of the pyramid to 2 significant figures. **[3 marks]**

EXAM TIP Find the area of the base first.

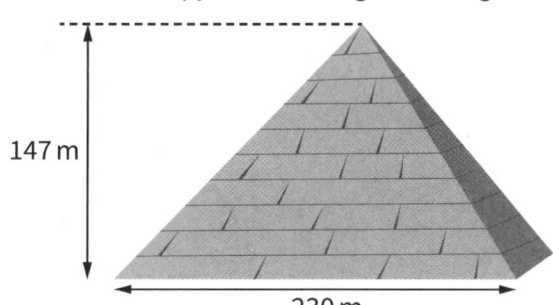

20.6 A sphere has a surface area of 400π cm². Work out the radius of the sphere. **[2 marks]**

EXAM TIP Form and solve an equation in r.

20.7 A thin hemispherical bowl has a radius of 25 cm.

Work out the maximum amount of water the bowl can hold.
Give your answer in millilitres to 3 significant figures. **[3 marks]**

20.8 Teresa models an ice cream cone out of plastic. The bottom cone part is a solid cone with radius 6 cm and height 15 cm. The top ice cream cone part is a solid hemisphere. The base of the hemisphere is the same size as the base of the cone.

Calculate, in terms of π, the volume of plastic in the model. **[3 marks]**

EXAM TIP Drawing a diagram might help here.

20.9 A fish tank is in the shape of a cuboid.
The tank has length 1 m, width 60 cm and height 80 cm.

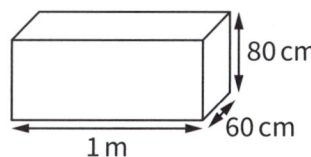

6 large jugs, each containing 1.5 litres of water, are poured into the tank.
The height of the water in the tank is now h cm.

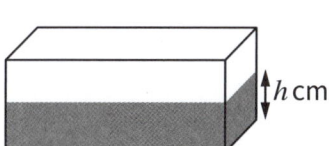

Work out the value of h. **[4 marks]**

Exam-style questions

20.10 Greta wants to cover the whole surface area of this closed cylinder in fabric. The cylinder has diameter 20 cm and height 40 cm. Greta has 3000 cm² of fabric.

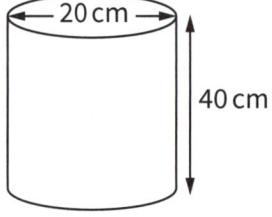

Work out whether or not Greta has enough fabric. Show all your working. [3 marks]

EXAM TIP
The surface area of a closed cylinder is given by the formula $2\pi rh + 2\pi r^2$.

20.11 The diagram shows a solid consisting of a hemisphere of diameter 12 cm on top of a cone with base diameter 12 cm and height h cm. The total volume of the solid is 276π cm³.

Work out the height of the cone, h. [4 marks]

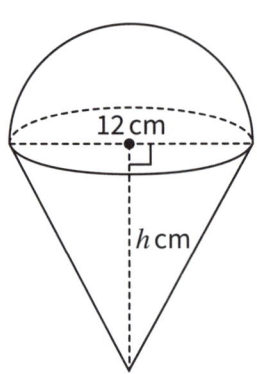

EXAM TIP
Volume of a cone
$= \dfrac{1}{3}\pi r^2 h$

Volume of a sphere
$= \dfrac{4}{3}\pi r^3$

Questions referring to previous content

20.12 At a post office, there are some medium parcels and large parcels. The parcels are to be posted either 1st or 2nd class.

The ratio of medium parcels to large parcels is 3 : 5.
The ratio of 1st class medium parcels to 2nd class medium parcels is 1 : 2.
The ratio of 1st class large parcels to 2nd class large parcels is 3 : 2.
What fraction of the parcels are posted 1st class? [4 marks]

20.13 Which circular sector has the greater area?
  A a 60° sector of a circle with radius 10 cm
  B a 75° sector of a circle with radius 9 cm [3 marks]

Knowledge 21

21 Pythagoras and 2D trigonometry

Pythagoras' theorem

Pythagoras' theorem for **right-angled triangles** is:

$a^2 + b^2 = c^2$

Where:
- c is the **hypotenuse** (longest side) of the triangle
- a and b are the other two sides.

You can use Pythagoras' theorem to find missing sides in right-angled triangles.

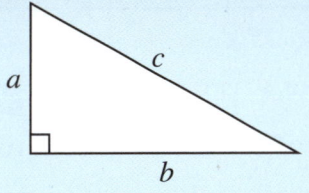

REVISION TIP

You can use Pythagoras' theorem to determine whether a triangle is right-angled.

If three sides of a triangle do not satisfy $a^2 + b^2 = c^2$, then the triangle is not right-angled.

Worked example

Calculate the length of the missing sides in these triangles.

1.

7 cm, 15 cm, x

$x^2 = 7^2 + 15^2$ — Substitute the known values into the formula.

$x = \sqrt{7^2 + 15^2}$ — Take the square root of both sides to find x.

$= 16.6$ cm (to 1 d.p.)

2.

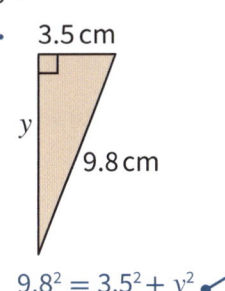

3.5 cm, 9.8 cm, y

$9.8^2 = 3.5^2 + y^2$ — Substitute the known values into the formula.

$y^2 = 9.8^2 - 3.5^2$ — Rearrange to make y the subject.

$y = \sqrt{9.8^2 - 3.5^2}$

$= 9.2$ cm (to 1 d.p.) — Then solve.

Worked example

The points A and B have coordinates $(1, 2)$ and $(4, -2)$. Calculate the length of AB.

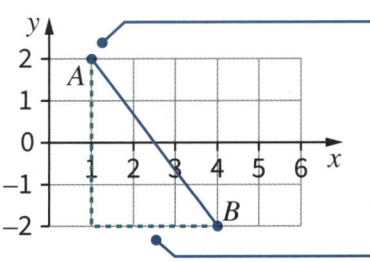

Plot the points A and B on a coordinate grid.

Make a right-angled triangle by drawing a vertical line from A and a horizontal line from B.

$(AB)^2 = 3^2 + 4^2 = 25$ — The triangle has a base of 3 and a height of 4.

$AB = \sqrt{25} = 5$

Knowledge

21 Pythagoras and 2D trigonometry

Pythagoras' theorem

Worked example

The base of a cone has radius 12 cm. Its volume is 768π cm³.

Calculate its surface area.

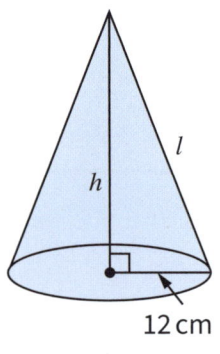

REVISION TIP

These formulae will be given to you in an exam:

surface area of cone $= \pi r l + \pi r^2$

volume of cone $= \frac{1}{3}\pi r^2 h$

$V = \frac{1}{3}\pi \times 12^2 \times h = 768\pi$

$48\pi h = 768\pi$

$h = 16$ cm

Slant height is needed to calculate surface area. First use the formula for the volume of a cone to find the perpendicular height.

Sketch a right-angled triangle inside the cone.

$l = \sqrt{16^2 + 12^2}$

$= 20$ cm

Use Pythagoras' theorem to find l.

curved surface area $= \pi \times r \times l$

$= \pi \times 12 \times 20$

$= 240\pi$ cm²

Find the curved surface area.

area of base $= \pi \times 12^2$

$= 144\pi$ cm²

Find the area of the base.

total surface area $= 240\pi + 144\pi$

$= 384\pi$

≈ 1206 cm²

Add together the areas to find the total surface area.

Trigonometry

Use **trigonometry** to find missing side lengths or angles in right-angled triangles.

Hypotenuse: The longest side of a right-angled triangle.

Opposite: Opposite the angle θ.

Adjacent: Next to the angle θ.

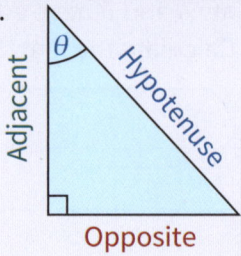

Formula

There are three trigonometric ratios that we can use to find missing sides or angles.

$\sin \theta = \dfrac{\text{opposite}}{\text{hypotenuse}}$

$\cos \theta = \dfrac{\text{adjacent}}{\text{hypotenuse}}$

$\tan \theta = \dfrac{\text{opposite}}{\text{adjacent}}$

REVISION TIP

Use this mnemonic to remember the trigonometric ratios.

SOH-CAH-TOA

Solving trigonometry problems

Sometimes you will be given two side lengths and asked to work out an angle.

The method is the same, even though you don't know what the angle is yet.

Worked example

Work out the value of x in this triangle.

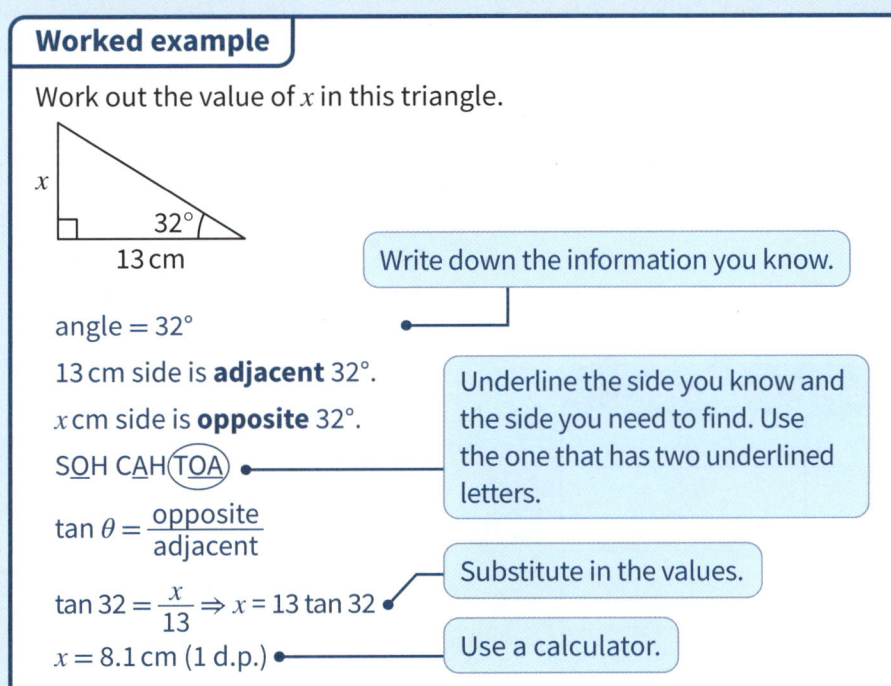

angle = 32°

13 cm side is **adjacent** 32°.

x cm side is **opposite** 32°.

S<u>O</u>H CA<u>H</u> <u>TO</u>A

$\tan \theta = \dfrac{\text{opposite}}{\text{adjacent}}$

Write down the information you know.

Underline the side you know and the side you need to find. Use the one that has two underlined letters.

$\tan 32 = \dfrac{x}{13} \Rightarrow x = 13 \tan 32$

Substitute in the values.

$x = 8.1$ cm (1 d.p.)

Use a calculator.

WATCH OUT

A lot of students lose marks because they don't know how to use their calculator properly.

Find these buttons:

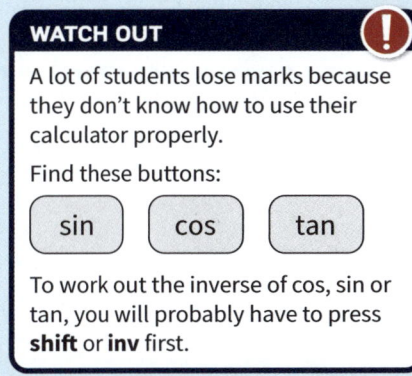

To work out the inverse of cos, sin or tan, you will probably have to press **shift** or **inv** first.

Worked example

A plank of wood is propped up against a wall. The plank rests 0.5 m from the base of the wall and makes an angle of 70° with the floor.

Work out the length of the plank to the nearest centimetre.

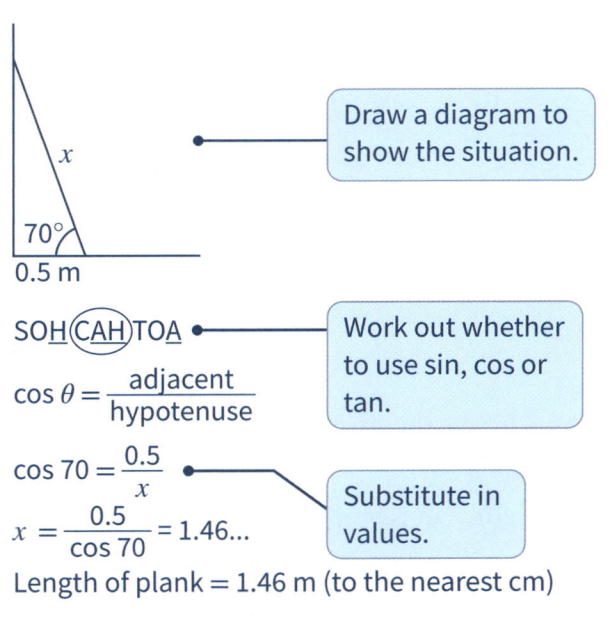

Draw a diagram to show the situation.

S<u>O</u>H <u>CAH</u> TOA

$\cos \theta = \dfrac{\text{adjacent}}{\text{hypotenuse}}$

Work out whether to use sin, cos or tan.

$\cos 70 = \dfrac{0.5}{x}$

$x = \dfrac{0.5}{\cos 70} = 1.46...$

Substitute in values.

Length of plank = 1.46 m (to the nearest cm)

Worked example

Work out the size of the angle marked x in this triangle.

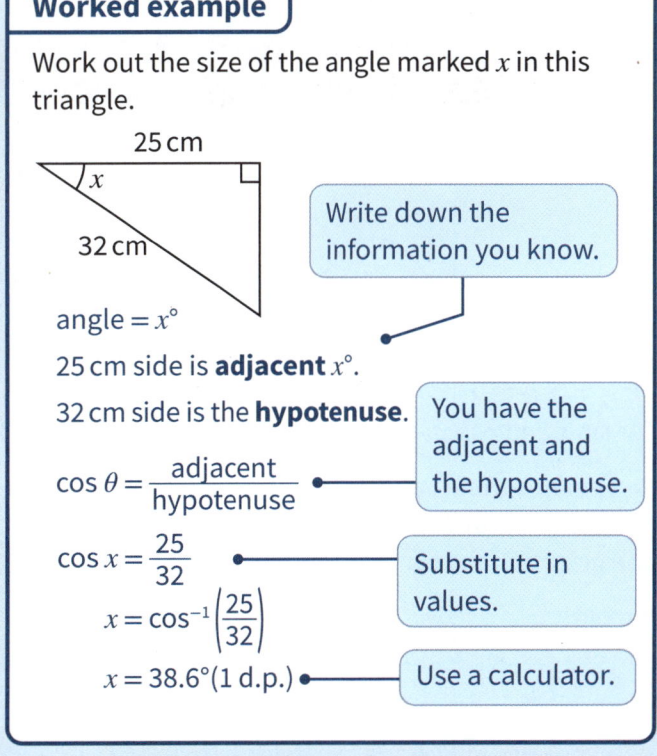

Write down the information you know.

angle = $x°$

25 cm side is **adjacent** $x°$.

32 cm side is the **hypotenuse**.

$\cos \theta = \dfrac{\text{adjacent}}{\text{hypotenuse}}$

You have the adjacent and the hypotenuse.

$\cos x = \dfrac{25}{32}$

Substitute in values.

$x = \cos^{-1}\left(\dfrac{25}{32}\right)$

$x = 38.6°$ (1 d.p.)

Use a calculator.

Knowledge

21 Pythagoras and 2D trigonometry

Trigonometric values

Here are the exact values you need to know (or be able to calculate quickly) in an exam. You can use the triangles or memorise the table.

First, draw a right-angled isosceles triangle with base and height of 1.

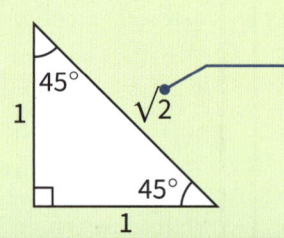

Use Pythagoras' theorem to find these lengths.

Now draw an equilateral triangle with side lengths 2 and split it into two right-angled triangles.

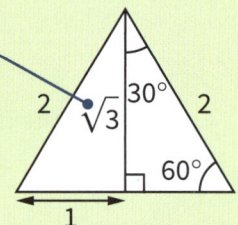

angle θ	$\sin \theta$	$\cos \theta$	$\tan \theta$
0°	0	1	0
30°	$\frac{1}{2}$	$\frac{\sqrt{3}}{2}$	$\frac{\sqrt{3}}{3}$
45°	$\frac{1}{\sqrt{2}} = \frac{\sqrt{2}}{2}$	$\frac{1}{\sqrt{2}} = \frac{\sqrt{2}}{2}$	1
60°	$\frac{\sqrt{3}}{2}$	$\frac{1}{2}$	$\sqrt{3}$
90°	1	0	n/a

Graphs of trigonometric functions

Learn the shapes and key features of these graphs.

$y = \sin x$

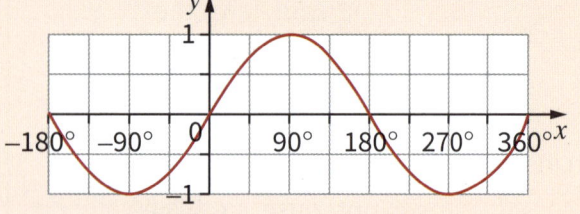

$y = \cos x$

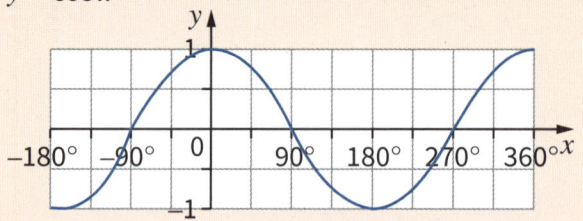

For $\sin x$ and $\cos x$:
- The maximum value is 1
- The minimum value is −1

The graphs of $y = \sin x$ and $y = \cos x$ both repeat every 360°.

$y = \tan x$

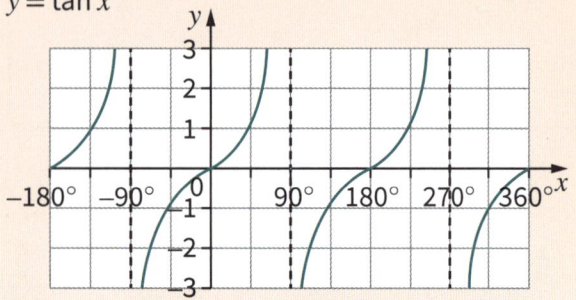

$\tan x$:
- does not have a maximum or minimum value.
- can be any value, but is undefined at $x = 90°, 270°, \ldots$

The graph of $y = \tan x$ repeats every 180°.

Graphs of trigonometric functions

Worked example

(a) Write the value of sin 60°.

$\sin 60° = \dfrac{\sqrt{3}}{2}$

(b) (i) Use your answer to part a) and the graph of $y = \sin x$ to work out the value of sin 120°.

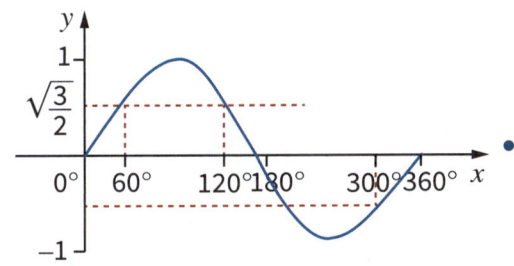

Sketch the sine graph between 0° and 360°.

$\sin 120° = \dfrac{\sqrt{3}}{2}$

(ii) Now find sin 300°.

$\sin 300° = -\sin 60°$

$= -\dfrac{\sqrt{3}}{2}$

Worked example

The graph shows $y = \sin x$.

On the same axes, sketch the graph of $y = \sin(x - 45°)$ for $0° \leq x \leq 360°$.

Label the coordinates of the points where $y = \sin(x - 45°)$ cuts the x-axis.

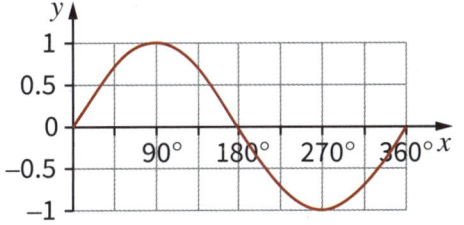

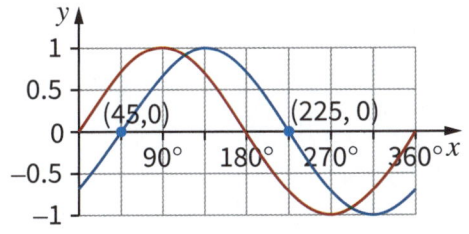

$y = \sin(x - 45°)$ is a translation of $y = \sin x$ by 45° in the direction of the positive x-axis. It cuts the x-axis at (45°, 0) and (225°, 0)

Key terms — Make sure you can write a definition for these key terms

adjacent   cosine (cos)   hypotenuse   opposite
Pythagoras' theorem   right-angled triangle
sine (sin)   tan   trigonometry

21 Knowledge

Retrieval

21 Pythagoras and 2D trigonometry

Learn the answers to the questions below, then cover the answers column with a piece of paper and write as many as you can. Check and repeat.

Questions / Answers

#	Question	Answer
1	What kind of triangles can Pythagoras' theorem be used with?	Right-angled triangles.
2	What is the hypotenuse?	The longest side of a right-angled triangle.
3	What is the value of sin 30°?	0.5
4	What is the value of tan 45°?	1
5	What is the formula used to find $\sin \theta$?	$\sin \theta = \dfrac{\text{opposite}}{\text{hypotenuse}}$
6	What is the formula used to find $\cos \theta$?	$\cos \theta = \dfrac{\text{adjacent}}{\text{hypotenuse}}$
7	Is $\dfrac{\text{opposite}}{\text{adjacent}}$ the formula for $\sin \theta$, $\cos \theta$ or $\tan \theta$?	$\tan \theta$
8	What is the value of cos 60°?	0.5
9	What is the exact value of sin 60°?	$\dfrac{\sqrt{3}}{2}$
10	If $\tan \theta = 5$, how do you find θ?	Use $\tan^{-1}(5)$
11	What is the value of cos 90°?	0
12	When finding the distance between two points on a grid, what theorem can you use?	Pythagoras' theorem.

Previous questions

Now go back and use these questions to check your knowledge of previous topics.

Questions / Answers

#	Question	Answer
1	What does LCM stand for?	Lowest common multiple.
2	What is the opposite of factorising?	Expanding.
3	When might you use the quadratic formula?	To solve a quadratic equation that cannot be factorised.
4	What do the letters BIDMAS stand for?	Brackets, indices, division, multiplication, addition and subtraction.

Practice 21

Exam-style questions

21.1 Only one of these triangles is right-angled. Which one? **[2 marks]**

21.2 Point A has coordinates (−1, 3). Point B has coordinates (2, 8).

Work out the length of the line segment AB to 3 significant figures. **[4 marks]**

EXAM TIP
A diagram is useful for this type of question.

21.3 The diagram shows the dimensions of a prize-winning biscuit in a baking competition. The biscuit is in the shape of an isosceles triangle.

Work out the height of the biscuit to 1 decimal place. **[3 marks]**

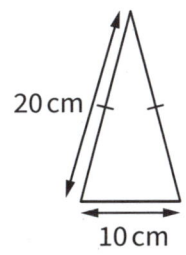

21.4 The diagram shows three identical steps. Work out the height h of each step to 1 decimal place. **[4 marks]**

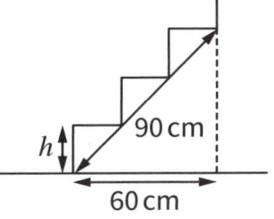

21.5 A solid cone has height 24 cm and radius 10 cm. Work out, giving your answers in terms of π,

(a) the volume of the cone **[2 marks]**

(b) the surface area of the cone. **[4 marks]**

EXAM TIP
The volume of a cone is given by $\frac{1}{3}\pi r^2 h$

The curved surface area of a cone is given by $\pi r l$.

21.6 Use your calculator to work out $\sqrt{\dfrac{\sin 47° \cos 21°}{\cos 21° - \sin 47°}}$

Give your answer correct to 3 significant figures. **[2 marks]**

21.7 Work out the length of the side labelled x in each of these triangles.
Give your answers to 1 decimal place: **[4 marks]**

(a)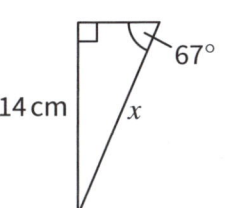

(b)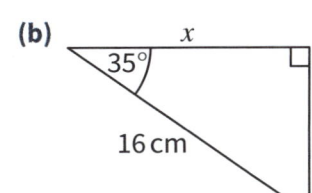

Exam-style questions

21.8 Work out the height *h* of this trapezium to 3 significant figures. [3 marks]

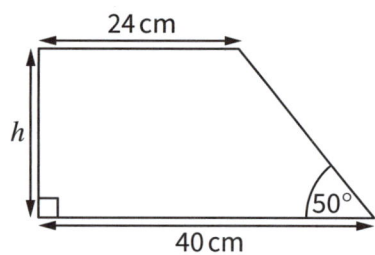

EXAM TIP
Split the trapezium up into separate shapes.

21.9 (a) Sketch the graph of $y = \cos x$ for $-360° \leq x \leq 360°$. [2 marks]

(b) Write the coordinates of the first minimum point on the graph where $x > 0$. [1 mark]

21.10 Work out the length of side *b* in the diagram. Give your answer to 1 decimal place. [3 marks]

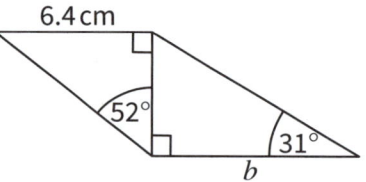

21.11 Work out the size of the angle *x* in each of these triangles. Write your answers to 1 decimal place. [6 marks]

(a) 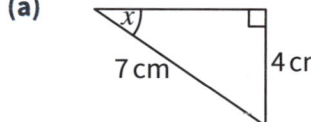 **(b)** 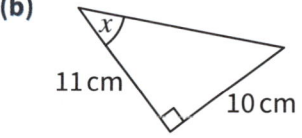 **(c)**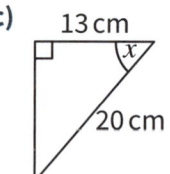

21.12 Triangle *PQR* is an isosceles triangle. Work out angle *PQR* to 1 decimal place. [3 marks]

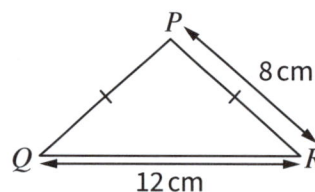

EXAM TIP
Split the isosceles triangle into two right-angled triangles.

21.13 *ABC* is a triangle. $AC = 7$ cm, $BC = 16$ cm and angle $BAC = 90°$.

Sketch the triangle and work out angle *BCA*. [3 marks]

21.14 (a) Sketch the graph of $y = \sin x$ for $0° \leq x \leq 450°$. [2 marks]

(b) The value of $\sin 45°$ is $\frac{1}{\sqrt{2}}$. Write all the other angles, $\theta°$, for which $\sin \theta = \frac{1}{\sqrt{2}}$ in the interval $0° \leq \theta \leq 450°$. [2 marks]

21.15 Work out the length of the side labelled x in each of these triangles. [9 marks]

(a) 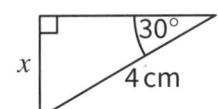 (b) 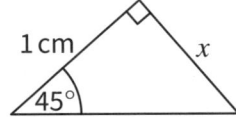 (c)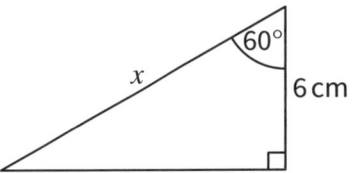

21.16 Arthur has completed a table of exact trigonometric ratios, but some of the values are incorrect.

	0°	30°	45°	60°	90°
$\sin \theta$	0	$\frac{1}{2}$	$\frac{1}{\sqrt{2}}$	$\frac{\sqrt{3}}{2}$	0
$\cos \theta$	1	$\frac{\sqrt{3}}{2}$	$\frac{\sqrt{3}}{2}$	$\frac{1}{2}$	0
$\tan \theta$	0	$\sqrt{3}$	1	$\frac{\sqrt{3}}{2}$	undefined

(a) Identify the incorrect values and circle them in the table. [4 marks]

(b) For each value that you have circled, clearly write the correct values below. [4 marks]

21.17 Work out the exact value of x in each triangle shown. [6 marks]

(a) 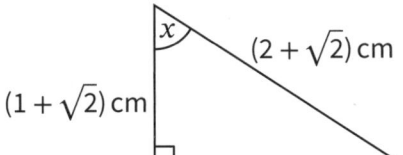   (b)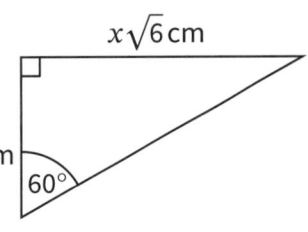

21.18 (a) Sketch the graph of $y = \tan x$ for $-180° \leq x \leq 360°$. [2 marks]

(b) Write the value of $\tan 30°$. [1 mark]

(c) Use the graph to work out the value of

(i) $\tan 210°$ [1 mark]

(ii) $\tan(-30°)$. [1 mark]

21 Practice

Exam-style questions

21.19 Sketch the graph of $y = \sin(x + 90)°$ for $0° \leq x \leq 360°$. [2 marks]

21.20 Sketch the graph of $y = -\cos x$ for $-180° \leq x \leq 180°$. [2 marks]

21.21 Tyrik says that there are no integers a, b and c such that the ratio $(\sin 45°)^2 : (\cos 30°)^2 : (\tan 30°)^2$ can be written in the form $a : b : c$.

Is Tyrik right? Show your working. [3 marks]

21.22 A square-based pyramid has a base length 12 cm and a height of 8 cm. F is the midpoint of CB.

Work out the length EF. [3 marks]

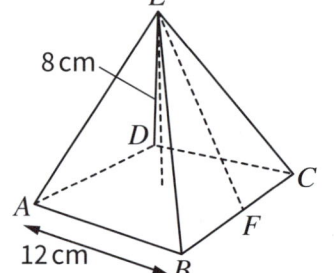

21.23 $ABCD$ is a trapezium.

Work out the size of angle ABC.
Give your answer to 1 decimal place. [5 marks]

21.24 ABC is a right-angled triangle.

(a) Work out the size of angle ACB.
Give your answer to 1 decimal place. [2 marks]

(b) The length of the side BC is increased. The length of side AB remains the same.

Will the value of tan ACB increase or decrease? You must give a reason for your answer. [1 mark]

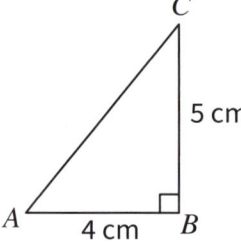

Questions referring to previous content

21.25 $x = 0.7\dot{0}\dot{2}$

Prove algebraically that x can be written as $\frac{26}{37}$. [3 marks]

21.26 The volume of a hemisphere is 60.75π cm^3

Find its surface area. What do you notice? [5 marks]

Knowledge 22

22 Similarity and congruence

Congruent shapes

Congruent shapes are exactly the same size and shape.
- Each pair of corresponding angles are equal.
- Each pair of corresponding sides are equal.
- The shapes may be reflected, rotated or translated, but the triangles are still congruent.

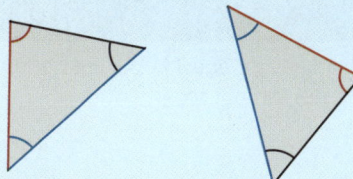

Corresponding sides are sides connecting the same angles on two shapes.

There are four conditions that mean two triangles are congruent.

The **three sides** are equal.
SSS (side–side-side)

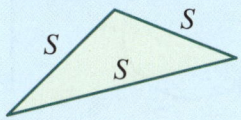 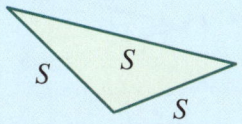

Two sides are equal and the **angle** between them are equal.
SAS (side – angle – side)

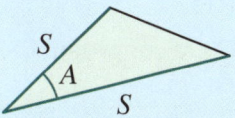 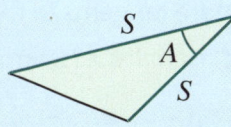

Two angles are equal, and one **side** are equal.
ASA (angle – side – angle)

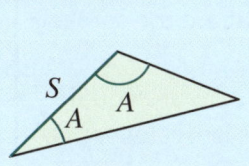 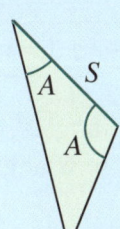

Both have a **right angle**; and the **hypotenuse** and one other **side** are equal.
RHS (right angle, hypotenuse, side)

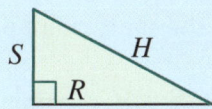 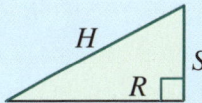

Worked example

Explain why the two triangles are congruent.

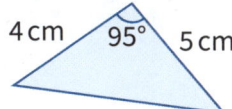 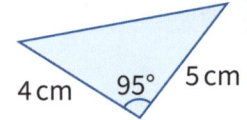

They are congruent because they satisfy the condition SAS.

$$\text{side 1} = 4\,\text{cm}$$
$$\text{angle between two sides} = 95°$$
$$\text{side 2} = 5\,\text{cm}$$

Worked example

Prove ABX and CDX are congruent.

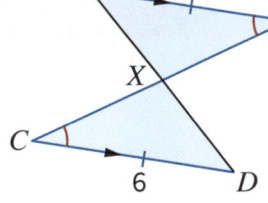

REVISION TIP

There are other ways you could prove that the triangles are congruent – this is just one!

angle DCB = angle ABC — These are pairs of alternate angles.
angle CDA = angle BAD
$CD = AB$ — Find the equal sides.

ABX and CDX are congruent because they satisfy the condition ASA. — State the condition the triangles satisfy.

Knowledge

22 Similarity and congruence

Congruent shapes

> **WATCH OUT**
>
> If only the three angles are equal (AAA), but not the side lengths, the triangles are **similar**, not congruent.

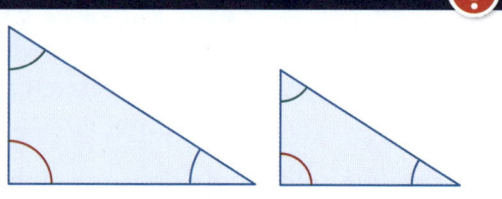

Worked example

A, B and C are three points on the circumference of a circle.

angle ABC = angle ACB

TB and TC are tangents to the circle.

Prove that triangles ATB and ATC are congruent.

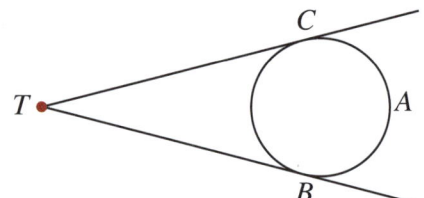

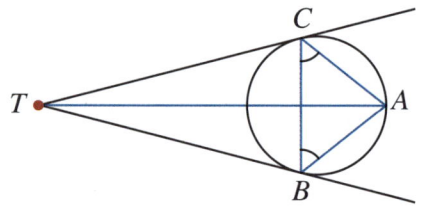

$TB = TC$ (tangents from the same point are equal) — Find the angles that are the same.

Annotate the image by drawing in lines to make a triangle, and marking any angles you know are the same.

$AB = AC$ — (angles ABC and ACB are equal, so triangle ABC is isosceles).

AT is a common side, so triangles ATB and ATC are congruent due to SSS. — State the condition the triangles satisfy.

Similar shapes

Similar shapes have exactly the same shape, but different sizes. One shape is an **enlargement** of the other.

The **scale factor** is the number you multiply side lengths of one shape by to get the side lengths of another shape.

$$\text{scale factor} = \frac{\text{new length}}{\text{original length}}$$

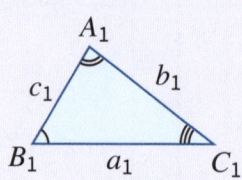
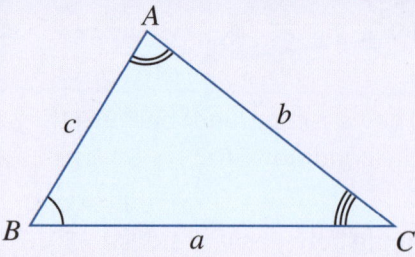

Each pair of corresponding angles are equal

Corresponding sides are in the same proportion:

$A = A_1, B = B_1, C = C_1$

multiply every side in 1st triangle by 2 to get the lengths in 2nd triangle:

$a = 2 \times a_1$
$b = 2 \times b_1$
$c = 2 \times c_1$

22

Similar shapes

Worked example

Triangles ABC and DEF are similar.

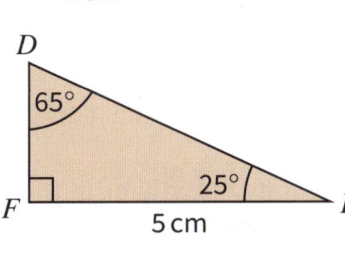

1. Calculate the length of DE.

	triangle ABC	triangle DEF
base	2	5
side	3	?

scale factor $= \dfrac{5}{2}$

length of $DE = 3 \times \dfrac{5}{2}$

$= 7.5$

- Make a table to show the corresponding values.
- Use the side that you know on each triangle to calculate the scale factor.
- Multiply the side corresponding to DE by the scale factor.

2. Write the size of angle CAB.

$\angle CAB = \angle FDE$

$\angle CAB = 65°$

- Find the corresponding angles.

REVISION TIP

scale factor for area = (scale factor for length)^2
scale factor for volume = (scale factor for length)^3

Worked example

The rectangles are similar. Find the length of side x.

8 cm, 12 cm

600 cm^2, x

$8 \times 12 = 96$ cm^2

$\dfrac{600}{96} = 6.25$ cm^2

$\sqrt{6.25} = 2.5$

12 cm $\times 2.5 = 30$ cm

- The area of the second shape has been given, so calculate the area of the first shape.
- Find the scale factor for the area. Square root the area scale factor to find the scale factor for length.
- Multiply the corresponding side length by the scale factor to find the length of side x.

A **frustum** is made by removing a small cone from the top of a larger cone.

LINK

To remind yourself about finding the volume of a cone, look back at Chapter 20.

REVISION TIP

For a cone with radius r, height h and slope height l:

volume of cone $= \dfrac{1}{3}\pi r^2 h$

surface area of cone $= \pi r l + \pi r^2$

These formulae will be given to you in an exam.

22 Knowledge

Knowledge

22 Similarity and congruence

Similar shapes

Worked example

Calculate the volume of the frustum in terms of π.

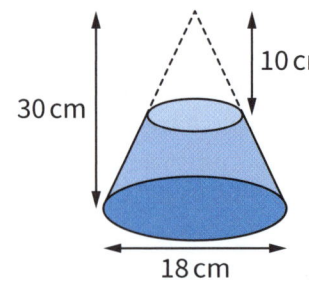

scale factor $= \dfrac{10}{30}$

$= \dfrac{1}{3}$ ← The cones are similar. Calculate the scale factor for the length.

radius of larger cone $= \dfrac{18}{2}$

$= 9$ cm ← Calculate the radius of the larger cone.

radius of smaller cone $= 9 \times \dfrac{1}{3}$

$= 3$ cm ← Use the scale factor to calculate the radius of the smaller cone.

volume of larger cone $= \dfrac{1}{3} \pi \times 9^2 \times 30$

$= 810\pi$ cm³ ← Calculate the volume of the larger cone in terms of π.

volume of smaller cone $= \dfrac{1}{3} \pi \times 3^2 \times 10$

$= 30\pi$ cm³ ← Calculate the volume of the smaller cone in terms of π.

volume of frustum $= 810\pi - 30\pi$

$= 780\pi$ cm³ ← Find the volume of the frustum.

Worked example

Two similar pyramids, A and B, have surface areas 425 cm² and 153 cm² respectively.

The volume of Pyramid B is 522 cm³.

Find the ratio of the volume of pyramid A to the volume of pyramid B.

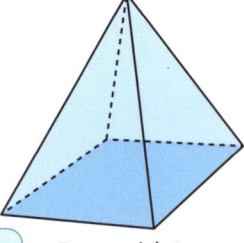 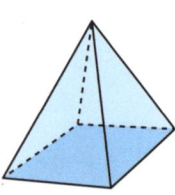

Pyramid A Pyramid B

$\dfrac{425}{153} = \dfrac{25}{9}$ ← Find the scale factor for the surface area.

$\sqrt{\dfrac{25}{9}} = \dfrac{5}{3}$ ← Take the square root to find the scale factor for the length.

$\left(\dfrac{5}{3}\right)^3 = \dfrac{125}{27}$ ← Cube your answer to find the scale factor for the volume.

$125 : 27$ ← Write the ratio.

Key terms Make sure you can write a definition for these key terms.

congruent shapes   corresponding sides   enlargement
frustum   scale factor   similar shapes

Retrieval

22 Similarity and congruence

Learn the answers to the questions below, then cover the answers column with a piece of paper and write as many as you can. Check and repeat.

Questions | Answers

#	Question	Answer
1	What does congruent mean?	Each pair of corresponding angles are equal. Each pair of corresponding sides are equal.
2	What does SSS mean?	Side-Side-Side: All corresponding sides are the same length.
3	What does ASA mean?	Angle-Side-Angle: A pair of corresponding sides are the same length, and two pairs of corresponding angles are equal.
4	What does RHS mean?	Right angle-Hypotenuse-Side: Both are right angled triangles. Hypotenuses are the same length and one other side is the same length in both triangles.
5	What does it mean for two shapes to be similar?	They are the same shape, but not the same size.
6	What makes two or more shapes similar?	Each pair of corresponding angles are equal. Corresponding sides are in the same proportion.
7	How do you calculate the scale factor?	scale factor = $\frac{\text{new length}}{\text{original length}}$
8	Are an object and its enlargement congruent or similar shapes?	They are similar shapes.
9	When using the SAS condition for congruence of triangles, which angles need to be equal?	The angles between the two pairs of equal sides.
10	Are all equilateral triangles similar? How do you know?	Yes, as all pairs of corresponding angles are 60°.

Previous questions

Now go back and use these questions to check your knowledge of previous topics.

Questions | Answers

#	Question	Answer
1	How do you convert from cm² to mm²?	Multiply by 100.
2	How do you multiply fractions?	Multiply the numerators and multiply the denominators.
3	What is an algebraic expression?	A collection of letters and numbers with no equals or inequality sign.
4	What is a quadratic equation?	An equation where the highest power of the variable is 2.

Practice

Exam-style questions

22.1 The two triangles *ABC* and *FED* are congruent.

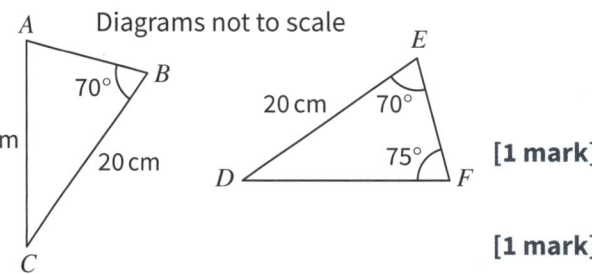

Diagrams not to scale

(a) Write the length of *DF*. [1 mark]

(b) Write the size of angle *CAB*. [1 mark]

22.2 Given that *AB* = *DE*, prove that triangles *CBA* and *CDE* are congruent. [4 marks]

EXAM TIP
Remember angle rules for parallel lines.

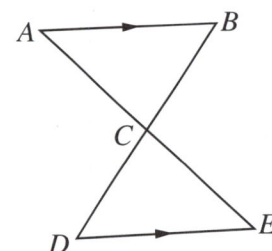

22.3 *ABCD* is a parallelogram.
JC is parallel to *AG*
BK is parallel to *ED*
Prove that triangles *ABG* and *CDE* are congruent. [4 marks]

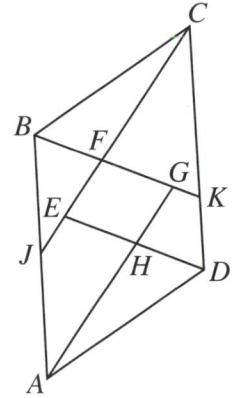

22.4 Prove that these two triangles are similar. [2 marks]

EXAM TIP
Compare ratios of sides.

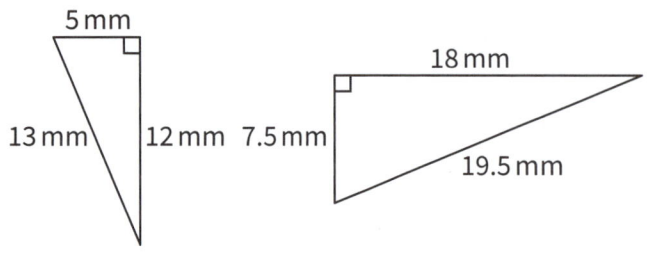

Diagrams not to scale

22.5 In the diagram, EB is parallel to DC.
ABE and ACD are similar triangles.
$AC = 11.5$ cm, $AB = 9.2$ cm
and $AE = 8.4$ cm.

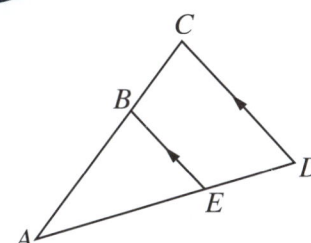

Work out the length of ED. **[2 marks]**

22.6 Bars A and B are similar cuboids. Bar A has length 10 cm and mass 1.5 kg.

Work out the mass of bar B given its length is 22 cm. **[3 marks]**

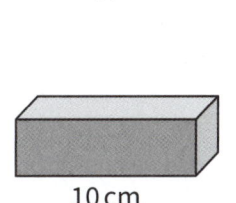 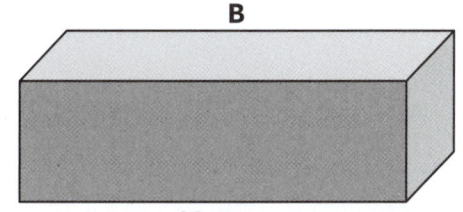

EXAM TIP

Mass is proportional to volume, so you should start by finding the volume scale factor.

22.7 Two mathematically similar shapes, A and B, are shown. The area of A is 12.5 cm² and the area of B is 50 cm². Shape A has a base of length 4 cm.

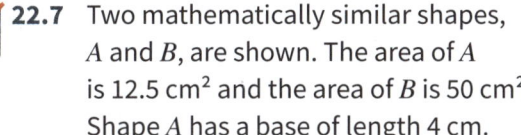

Work out the base length of shape B. **[3 marks]**

22.8 Two solids are similar. The larger solid has surface area 360 cm² and volume 675 cm³. The smaller solid has volume 25 cm³.

Work out the surface area of the smaller solid. **[3 marks]**

22.9 The diagram consists of three overlapping trapezia (large, medium, small) with the same centre.
The two parallel sides of the large trapezium measure 80 cm and 120 cm.
The medium trapezium is an enlargement of the large trapezium with scale factor 0.75.
The small trapezium is an enlargement of the medium trapezium with scale factor 0.5. The area of the shaded region is 4050 cm².

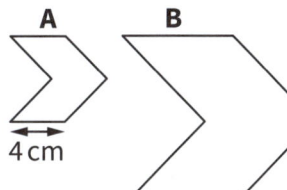

Work out the height, h, of the large trapezium. **[5 marks]**

Exam-style questions

22.10 A cork bottle-stopper is in the form of a frustum of a cone. The dimensions of the stopper are as shown in the diagram.

Work out, in terms of π, the volume of cork required to make the stopper. [5 marks]

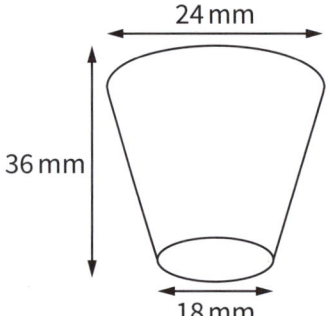

EXAM TIP

Remember, the cut-off cone is similar to the large cone.

22.11 Here are three tetrahedra.

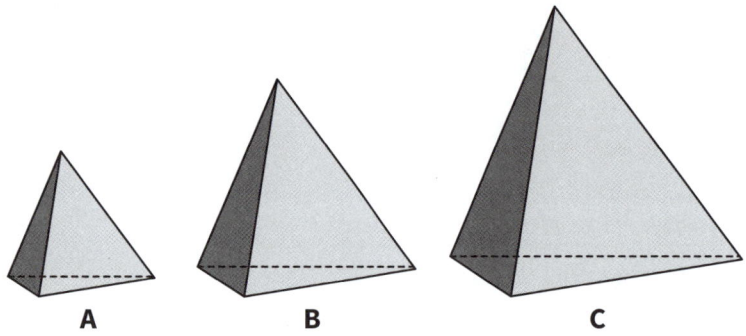

The volume of tetrahedron B is 20% greater than the volume of tetrahedron A.

The volume of tetrahedron B is 60% less than than the volume of tetrahedron C.

Find the volume of tetrahedron A as a fraction as the volume of tetrahedron C. Write your fraction in its simplest form. [3 marks]

22.12 Sphere J and sphere K are mathematically similar. The ratio of the volume of sphere J to the volume of sphere K is 125 : 8.

The surface area of sphere K is 460 cm².

Show that the surface area of sphere J is 2875 cm². [4 marks]

Questions referring to previous content

22.13 *ABC* is an isosceles triangle with
AC = *BC* = 15 cm and *AB* = 6 cm.

Work out the size of angle *CAB*, giving your
answer to 1 decimal place.　　　　　　　　　　　**[3 marks]**

22.14 The length of a rectangle is twice the width.
The area of the rectangle is 20 cm².

Work out the length of the rectangle.　　　　　　**[4 marks]**

Knowledge

23 Transformations

Transformations

A **transformation** is something that alters the size of a shape or its position.

- The original shape is called the **object**.
- The transformed shape is called the **image**.

Reflection

When a shape has been reflected, the object and image are **congruent**, but they do not have the same orientation.

Worked example

Describe the transformation that maps ABC to DEF.

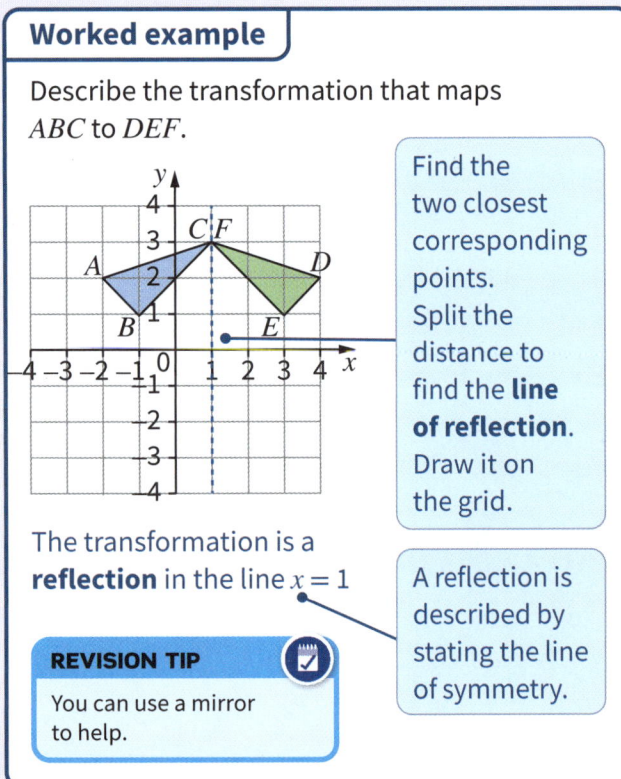

The transformation is a **reflection** in the line $x = 1$.

Find the two closest corresponding points. Split the distance to find the **line of reflection**. Draw it on the grid.

A reflection is described by stating the line of symmetry.

REVISION TIP
You can use a mirror to help.

Rotation

When a shape has been reflected, the object and image are **congruent**, but they do not have the same orientation.

Worked example

Describe the transformation that maps ABC to DEF.

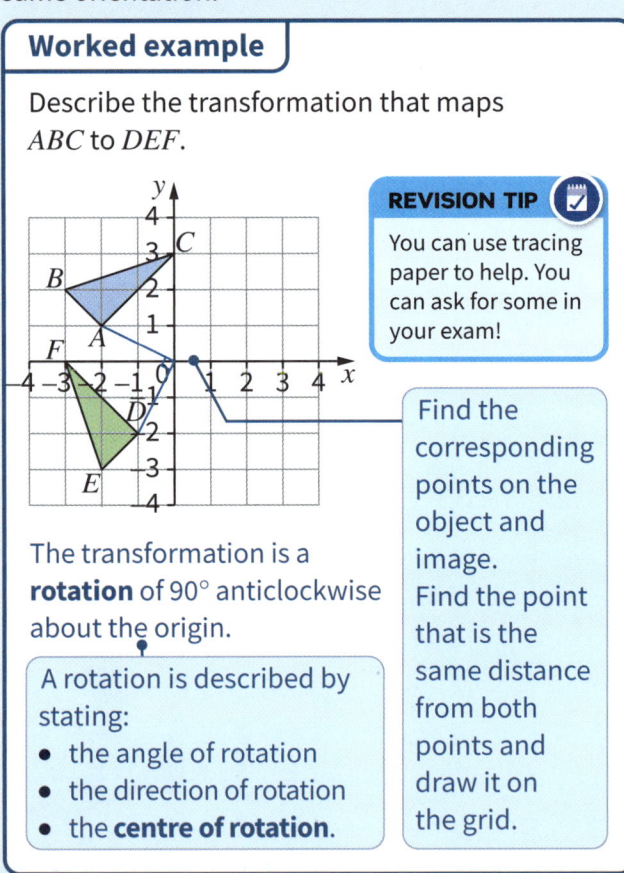

REVISION TIP
You can use tracing paper to help. You can ask for some in your exam!

The transformation is a **rotation** of 90° anticlockwise about the origin.

A rotation is described by stating:
- the angle of rotation
- the direction of rotation
- the **centre of rotation**.

Find the corresponding points on the object and image. Find the point that is the same distance from both points and draw it on the grid.

Transformations

A **translation** can be described using a **column vector**.

- The top number shows the movement in the x-axis.
- The bottom number shows the movement in the y-axis.

When a shape has been translated, the object and image are **congruent** and they have the same orientation (unlike rotations and reflections).

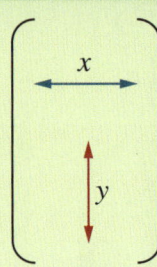

23

Translation

Worked example

1. What is the transformation that maps *ABC* to *DEF*?

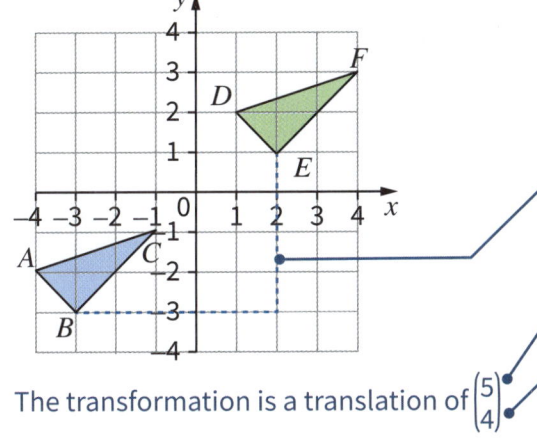

The transformation is a translation of $\begin{pmatrix} 5 \\ 4 \end{pmatrix}$.

- Starting from the object, choose a point and count the horizontal and vertical squares to the corresponding point on the image.

- The shape has moved:
  5 squares in the positive *x*-direction
  4 squares in the positive *y*-direction

2. What is the transformation that maps *DEF* to *ABC*?

The transformation is a translation of $\begin{pmatrix} -5 \\ -4 \end{pmatrix}$.

- To go from the image to the object, the translation is the inverse of the original translation. You can also find this by counting squares on the grid again.

Enlargement

To enlarge a shape, you need to know:
- the **centre of enlargement**
- the **scale factor**.

If the scale factor is:
- positive, the image will be on the same side of the centre of enlargement as the object, and will have the same orientation as the object
- negative, the image will be on the other side of the centre of enlargement, and will be reversed from the object
- greater than 1, the image will be bigger than the object
- less than 1 and greater than −1, the image will be smaller than the object.

For example, an object that is enlarged by $-\frac{3}{4}$ will have an image that is smaller than the object, and will be reversed, on the other side of the centre of orientation.

When a shape has been enlarged, the object and image will be **similar**, not congruent.

Key terms — Make sure you can write a definition for each of these key terms

centre of enlargement   centre of rotation   column vector
congruent   image   line of reflection   object
reflection   rotation   scale factor   similar
transformation   translation

23 Knowledge 181

Knowledge

23 Transformations

Enlargement

Worked example

Enlarge triangle ABC by scale factor 2 with centre of enlargement $(0, 0)$.

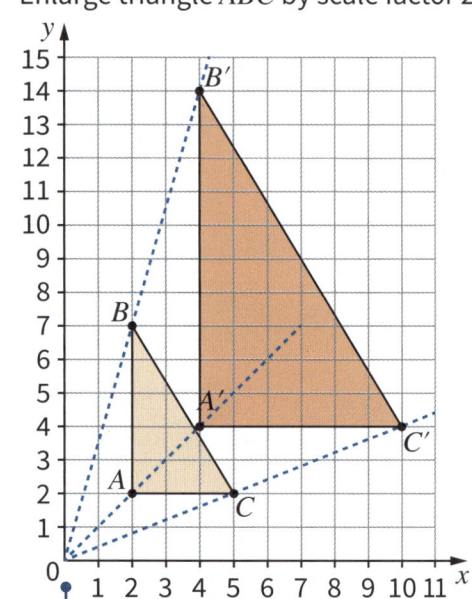

2 right; **2** up.

$2 \times 2 = \mathbf{4}; 2 \times 2 = \mathbf{4}$

$(0, 0) \Rightarrow (\mathbf{4, 4})$

$A' = (4, 4)$

$B' = (4, 7)$

$C' = (10, 4)$

- Measure the distance of vertex A from the centre of enlargement.
- Multiply these numbers by the scale factor.
- Add these numbers to the centre of enlargement to find vertex A'.
- Repeat the process for B and C to plot vertices B' and C'.
- centre of enlargement

Worked example

Enlarge the shape by scale factor 0.5 from centre of enlargement $(1, 1)$.

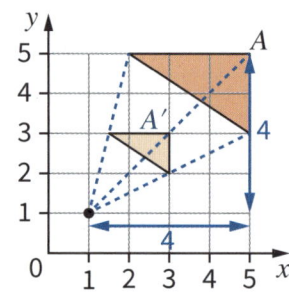

4 right, 4 up

$4 \times 0.5 = 2$

$4 \times 0.5 = 2$

$(1, 1) \rightarrow (3, 3) = A'$

- Measure the distance of vertex A from the centre of enlargement.
- Multiply by the scale factor.
- Add to centre of enlargement to find A'.

Worked example

Enlarge triangle A by scale factor -2, centre of enlargement $(0, 0)$.

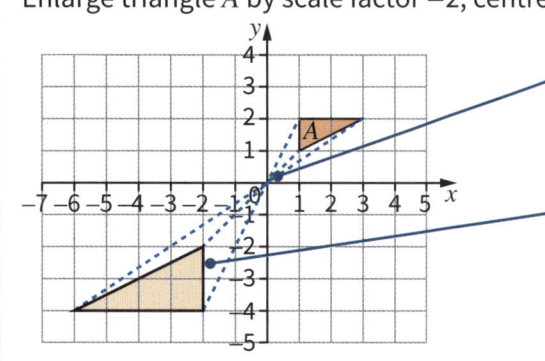

- Draw dotted lines from each of the vertices of the triangle through the centre of enlargement.
- The distance from the centre of enlargement to each of the vertices is multiplied by 2, but the image is on the opposite side of the centre of enlargement.

Retrieval

23 Transformations

Learn the answers to the questions below, then cover the answers column with a piece of paper and write as many as you can. Check and repeat.

Questions | Answers

1. What are the four types of transformation? — Reflection, rotation, translation, enlargement.
2. What do you need to know to draw the reflection of a shape on a graph? — The equation of the line of symmetry.
3. When you reflect a shape, are the object and the image congruent or similar? — Congruent.
4. What three pieces of information do you need to know to describe the rotation of a shape? — The angle, the direction and the centre of rotation.
5. What can you ask for in an exam to help with drawing rotations? — Tracing paper.
6. How do you write a translation mathematically? — Using a column vector.
7. When you translate a shape, are the object and image congruent or similar? — Congruent.
8. What does the top number of a column vector tell you? — How far to move right (if positive) or left (if negative).
9. To enlarge a shape, what two pieces of information do you need? — The scale factor and the centre of enlargement.
10. If you are given the object and the image, how do you find the centre of enlargement? — Draw lines through the corresponding vertices and see where they meet. This point is the centre of enlargement.

Previous questions

Now go back and use these questions to check your knowledge of previous topics.

Questions | Answers

1. What is a regular shape? — A shape that has all sides the same length and all angles the same size.
2. What is the formula that is used to calculate the sum of the interior angles in a polygon? — sum of interior angles = (number of sides − 2) × 180°
3. What can you say about the size of two corresponding angles on parallel lines? — They are equal.
4. What is the area of a 2D shape? — A measure of the space inside a 2D shape.
5. How do you find the area of a compound shape? — Split it into simpler shapes, find the area of each shape and then add the areas up.

Practice

Exam-style questions

23.1

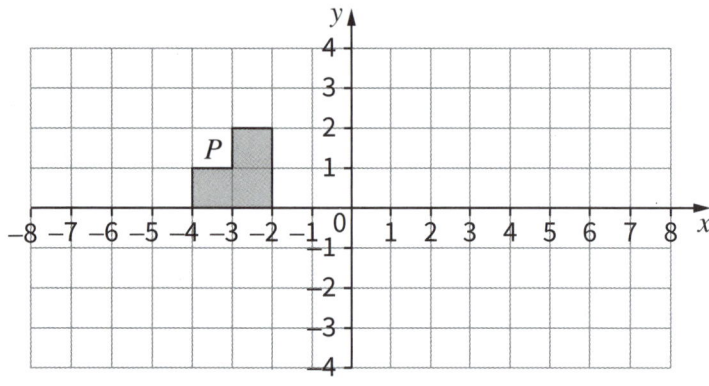

(a) Rotate shape P by 180° about the point $(1, 0)$.
Label the new shape A. **[1 mark]**

(b) Translate the shape P by the vector $\begin{pmatrix} -2 \\ -3 \end{pmatrix}$.
Label the new shape B. **[1 mark]**

23.2 Describe the following transformations fully.

> **EXAM TIP**
> If you are unsure, a piece of tracing paper may help you.

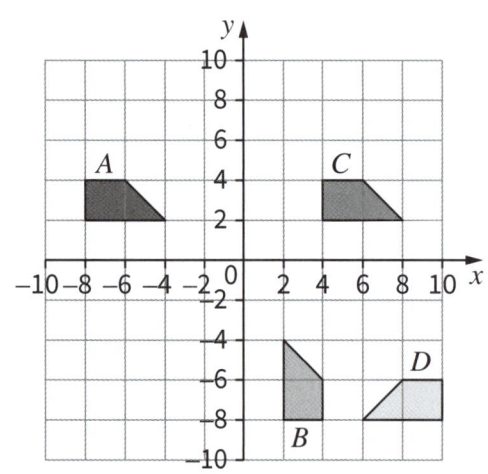

(a) A to C (b) B to A (c) D to B **[7 marks]**

23.3 (a) Use the diagram shown for this question.

(i) Reflect triangle P in the line $y = 0$.
Label the image Q. **[2 marks]**

(ii) Rotate triangle P 90° anticlockwise about $(-2, -2)$. Label the image R. **[2 marks]**

(iii) Translate triangle P by the vector $\begin{pmatrix} -8 \\ 4 \end{pmatrix}$.
Label the image S. **[1 mark]**

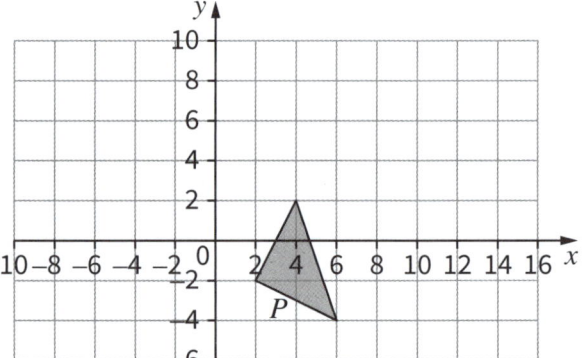

(b) Another triangle, T, has vertices with coordinates $(10, 8)$, $(12, 2)$ and $(14, 6)$. Describe the single transformation that maps triangle T onto triangle P. **[3 marks]**

> **EXAM TIP**
> For part **b**, start by drawing triangle T.

23.4 **(a)** Enlarge shape A with scale factor $\frac{1}{2}$ about the centre of enlargement $(0, 0)$. Label your image B. **[2 marks]**

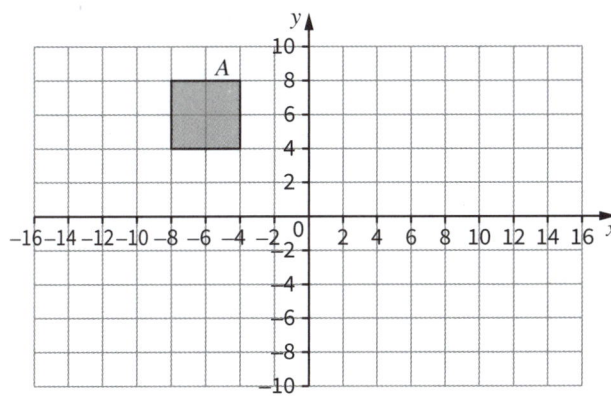

(b) On the same grid, enlarge shape A with scale factor -1 about the point $(0, 0)$. Label your image C. **[2 marks]**

(c) Write two other transformations that map shape A directly onto shape C. **[3 marks]**

23.5 Rotate shape P 90° clockwise about the point $(-6, 3)$ and then enlarge it with scale factor -2 about the point $(-2, 0)$. Label your final image Q. **[4 marks]**

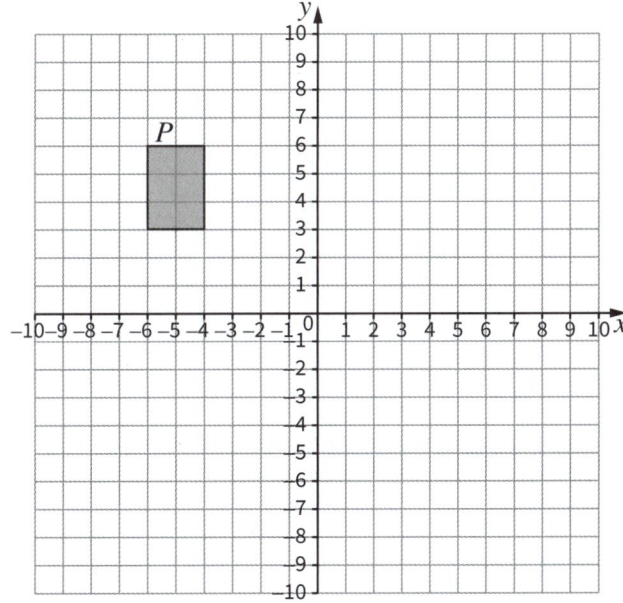

23.6 Describe the single transformation that maps triangle A onto triangle B.

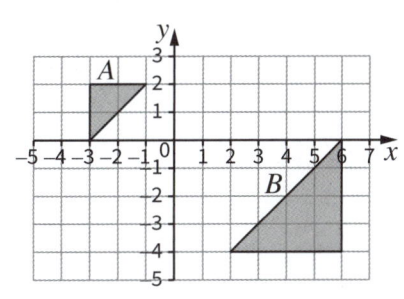

[3 marks]

Exam-style questions

23.7 Triangle P is reflected in the line $x = 2$ to give triangle Q. Triangle Q is reflected in the line $y = 0$ to give triangle R.

Describe the single transformation that maps triangle P onto triangle R. **[3 marks]**

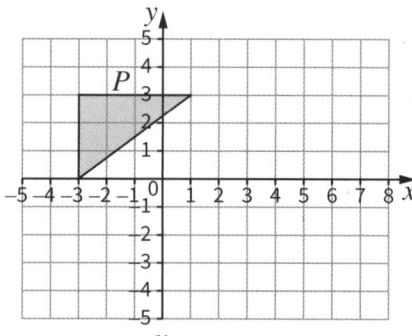

23.8 Rectangle D is transformed by the combined transformation of an anticlockwise rotation of 90° about $(4, -2)$ followed by a translation with vector $\begin{pmatrix} 8 \\ 4 \end{pmatrix}$.

One point on rectangle D is invariant under the combined transformation. Find the coordinates of the point. **[2 marks]**

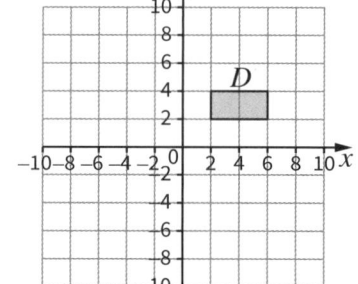

23.9 Enlarge shape A by scale factor $\frac{1}{3}$ about centre of enlargement $(4, 1)$.

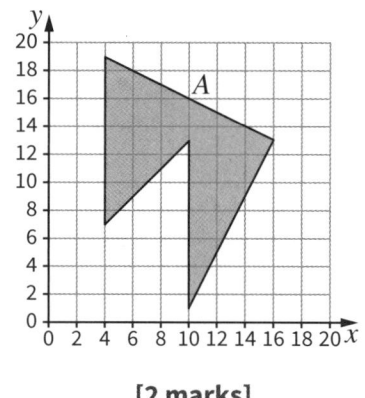

[2 marks]

Questions referring to previous content

23.10 A line, L, has equation $y = 4 - 3x$. Sajid says that the line $3x + y = 0$ is parallel to L. Is Sajid correct? Explain your reasoning. **[2 marks]**

23.11 100 worms are measured. The table shows their measurements in cm.

Length (x cm)	Frequency, f	Midpoint	$f \times$ midpoint
$0 < x \leq 8$	50	4	200
$8 < x \leq 16$	30	12	
$16 < x \leq 24$	20		400
Total	100	–	

(a) Complete the missing values in the table. **[2 marks]**

(b) Write the modal class. **[1 mark]**

(c) Work out an estimate for the mean length. **[2 marks]**

Knowledge 24

24 Plans, elevations, constructions, bearings

Plans and elevations

A **plan** is the view from directly above an object.

An **elevation** is the view from the side or the front of an object.

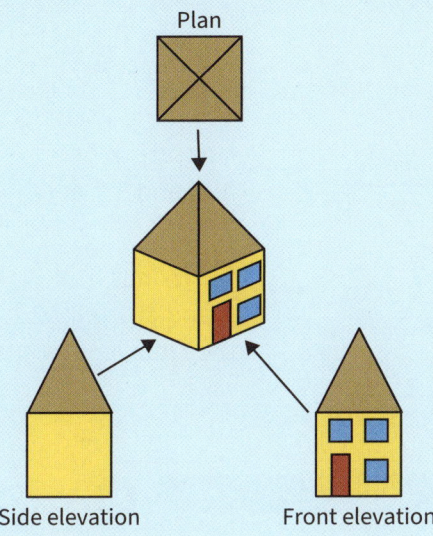

Worked example

The 3D solid is made using eight centimetre cubes.

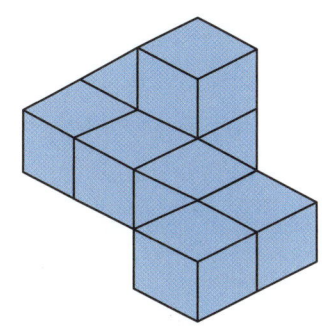

Draw a plan of the solid on a grid.

Start by drawing the outline of the lower layer, then add in lines to show the upper layer.

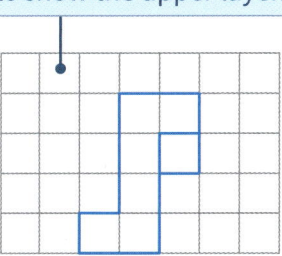

REVISION TIP

The plan can be drawn in any orientation, as long as it's from above.

Worked example

The plan view and the front and side elevations of a shape are shown.

| Plan | Front elevation | Side elevation |

The shape is made up of identical cubes. Draw a 3D sketch of the shape.

The plan shows the base of the shape is a rectangle of 3 by 2 cubes, with the top left cube raised.

The front and side elevations both show that the top left part of the plan view is raised by a height of 1 cube.

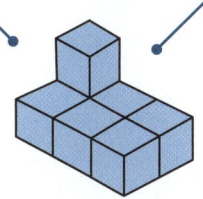

Constructing triangles

You can use a ruler and a protractor to accurately draw a triangle when you know one of these combinations:

- angle – side – angle (ASA)
- side – angle – side (SAS)
- hypotenuse and one other side of a right-angled triangle (RHS).

Knowledge

24 Plans, elevations, constructions, bearings

Constructing triangles

Worked example

In the triangle ABC, $AB = 6$ cm, $BC = 8$ cm and angle $ABC = 62°$.
Draw the triangle ABC.

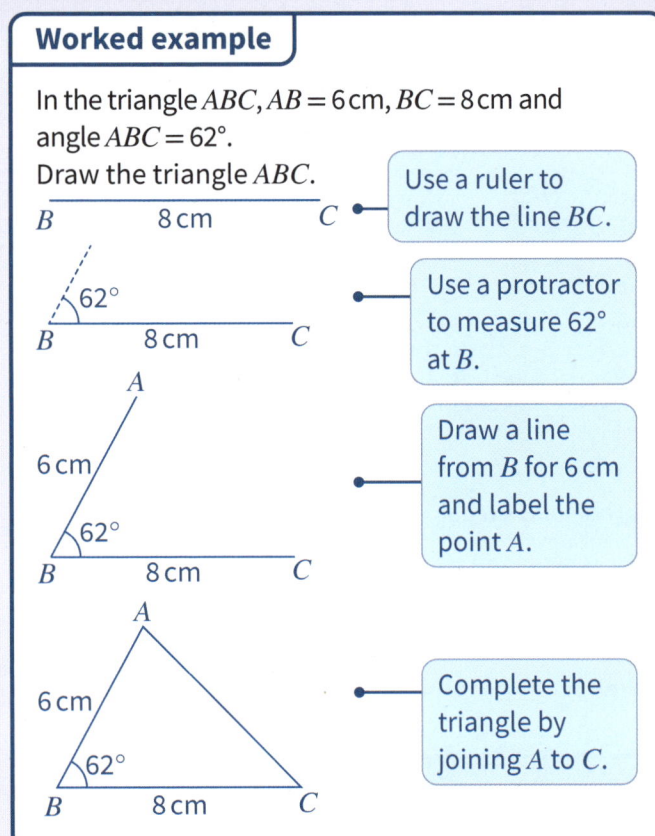

- Use a ruler to draw the line BC.
- Use a protractor to measure 62° at B.
- Draw a line from B for 6 cm and label the point A.
- Complete the triangle by joining A to C.

Worked example

In the triangle PQR, angle $RPQ = 43°$, angle $RQP = 57°$ and $PQ = 7$ cm.
Draw triangle PQR.

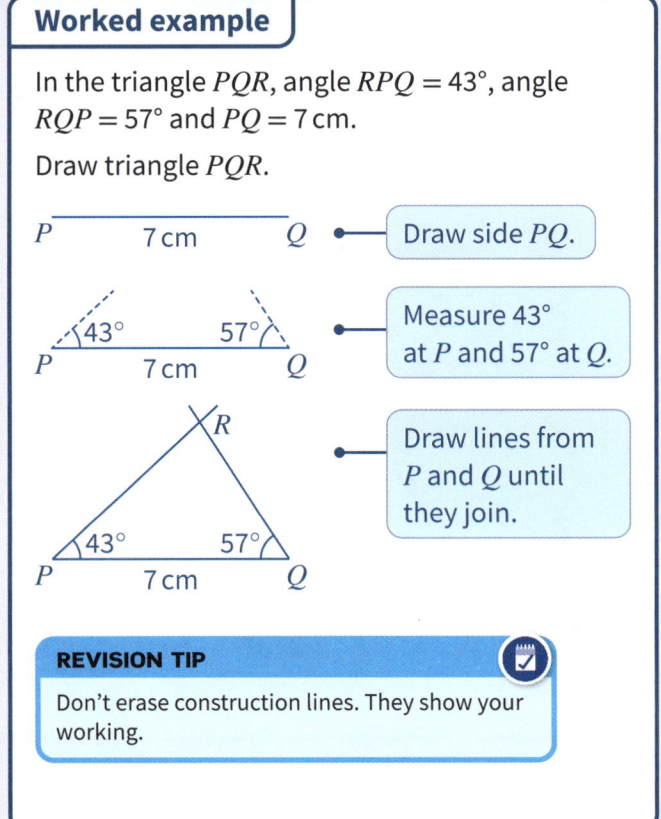

- Draw side PQ.
- Measure 43° at P and 57° at Q.
- Draw lines from P and Q until they join.

REVISION TIP

Don't erase construction lines. They show your working.

You can also use a ruler and a **pair of compasses** to **construct** a triangle if you know all three side lengths.

Worked example

Construct a triangle with side lengths 7 cm, 4 cm and 5 cm.

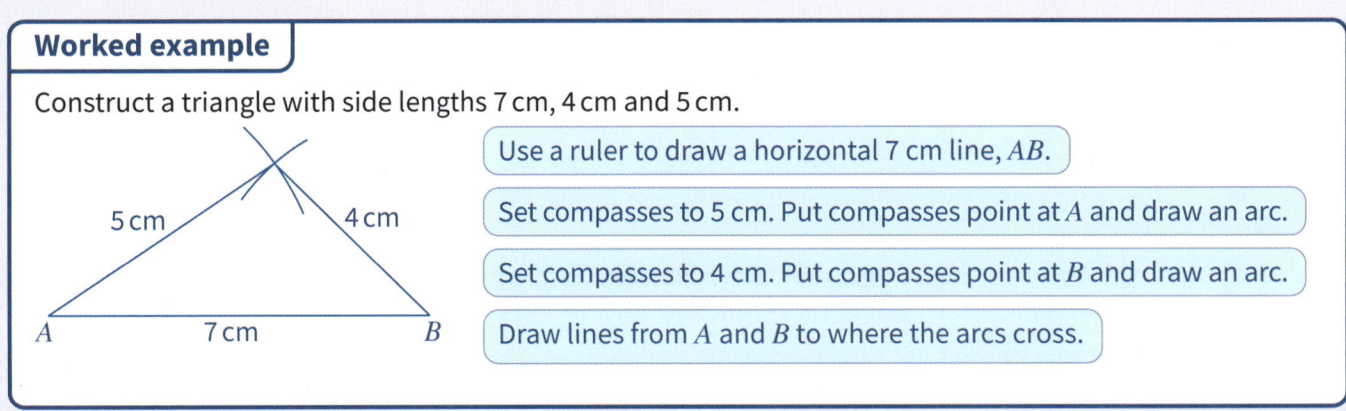

- Use a ruler to draw a horizontal 7 cm line, AB.
- Set compasses to 5 cm. Put compasses point at A and draw an arc.
- Set compasses to 4 cm. Put compasses point at B and draw an arc.
- Draw lines from A and B to where the arcs cross.

Constructing bisectors

Perpendicular lines are at 90° to each other (at right angles).

A perpendicular **bisector** cuts a line exactly in half at right angles. It can therefore be used to find the midpoint of a line.

The perpendicular distance is the shortest distance from a point to a line.

Perpendicular bisector

Worked example

Draw the perpendicular from the dot though the line.

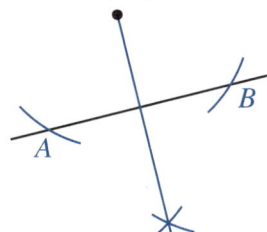

- Put compasses on the dot. Draw two arcs on the line. Label the intercepts A and B.
- Put compasses at A. Draw an **arc** below the line.
- Put compasses at B. Draw an arc below the line.
- Draw the line from the dot through where the arcs meet.

Worked example

Construct an angle of 135°

If the angle from the horizontal to a line is 135°, clockwise, then the angle from the horizontal to the same line, anticlockwise, is $180 - 135 = 45°$

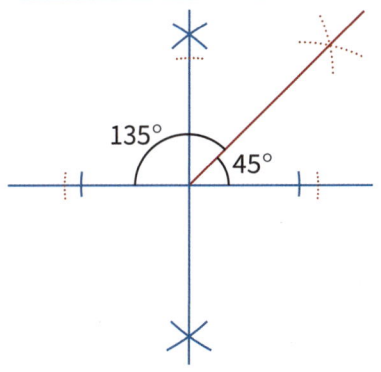

An angle of 45° is the bisector of a right angle. So:

Create the perpendicular bisector (see **blue** arcs and line).

Create the angle bisector. (see **red dotted** arcs and line).

WATCH OUT
Remember not to delete your construction lines - draw them clearly.

Angles can also be bisected. A bisector also cuts an angle exactly in half.

Worked example

Use compasses to bisect the angle.

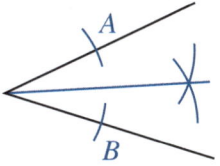

- Put compasses on the vertex. Draw an arc on each line. Label the intercepts A and B.
- Put compasses at A. Draw an arc to the right of the angle.
- Put compasses at B. Draw an arc to the right of the angle.
- Draw the line from the angle through where the arcs meet.

Knowledge

24 Plans, elevations, constructions, bearings

Loci

Locus (plural **loci**) are points that are all the same distance from a point or line.

The **locus of points** a fixed distance from a point is a circle.

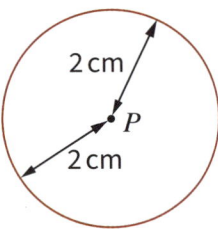

All the points that are 2 cm from P lie on the red circle.

The locus of points a fixed distance from a line is a pair of parallel lines..

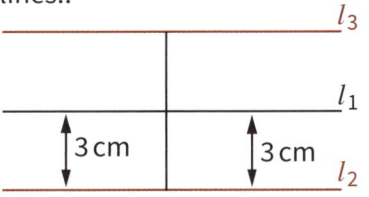

All the points that are 3 cm from l_1 lie on the red lines, l_2 and l_3.

The locus of points **equidistant** from two points is a perpendicular bisector of those two points.

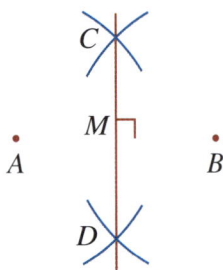

All the points that are equidistant from both A and B lie on the line CD.

The locus of points equidistant from two lines that intersect at a point is a bisector of the angle between the lines.

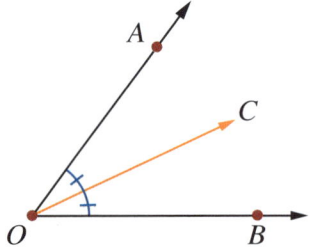

All the points that are **equidistant** from both OA and OB lie on the line OC.

Worked example

A runway is 3 km long. A fence surrounds the runway at a distance of 1.2 km the whole way around.

1. Construct a **scale drawing** of the runway, modelled as a line AB, and the fence, using a scale of 1 : 60 000

$$3 \text{ km} = 300\,000 \text{ cm}$$
$$1.2 \text{ km} = 120\,000 \text{ cm}$$

Convert the distances from km to cm.

length of AB in drawing
$$= \frac{300\,000}{60\,000}$$
$$= 5 \text{ cm}$$

Divide the lengths by the scale given to find how long the distances should be.

distance of fence in drawing
$$= \frac{120\,000}{60\,000}$$
$$= 2 \text{ cm}$$

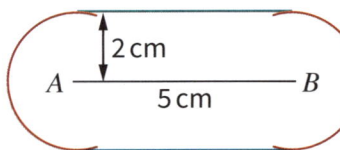

Draw a 5 cm line for the runway, then draw parallel lines 2 cm away and a semicircle of radius 2 cm at each end for the fence.

24

Loci

Worked example

2. Shade the region where the points inside the fence are closer to B than to A.

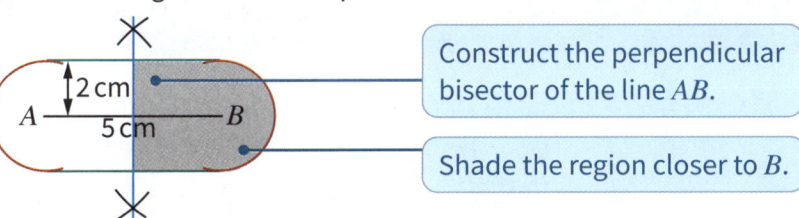

- Construct the perpendicular bisector of the line AB.
- Shade the region closer to B.

Bearings

A **compass** (not a pair of compasses!) shows the direction of magnetic north and the directions east, south and west.

A **bearing** indicates the direction of travel.

Bearings are angles that are measured clockwise from north.

Bearings always have three figures; for example, a bearing of 3° is written as 003°.

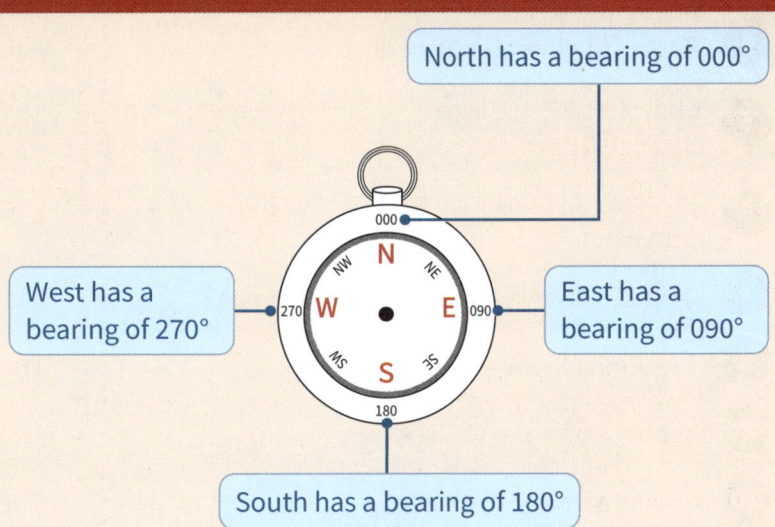

- North has a bearing of 000°
- West has a bearing of 270°
- East has a bearing of 090°
- South has a bearing of 180°

Worked example

A boat sails south-east from a rock.

1. What is the bearing of the boat from the rock?

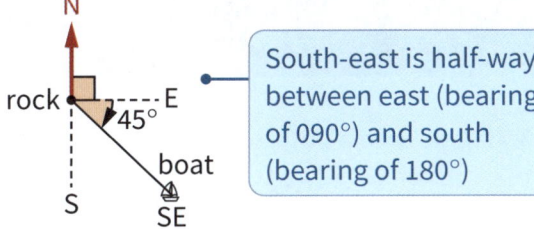

South-east is half-way between east (bearing of 090°) and south (bearing of 180°)

The bearing of the boat from the rock is 90° + 45° = 135°.

2. What is the bearing of the rock from the boat?

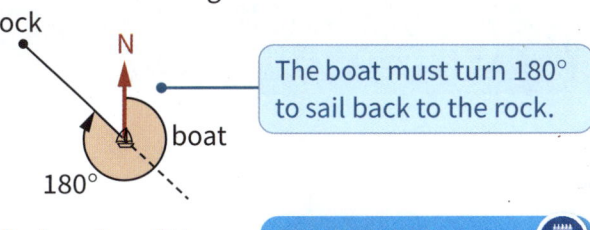

The boat must turn 180° to sail back to the rock.

The bearing of the rock from the boat is 135° + 180° = 315°.

REVISION TIP

Always draw a diagram!

LINK

To remind yourself about measuring angles in degrees look back at Chapter 19.

Key terms — Make sure you can write a definition for these key terms

arc   bearing   bisector   compass
construct   due (north, south, east, west)
equidistant   front elevation   locus
pair of compasses   perpendicular   plan
scale drawing   side elevation

24 Knowledge 191

Retrieval

24 Plans, elevations, constructions, bearings

Learn the answers to the questions below, then cover the answers column with a piece of paper and write as many as you can. Check and repeat.

	Questions	Answers
1	What is the first step for drawing a perpendicular bisector of a line segment?	Put the point of the compasses at one end of the line. Open the compasses to more than half the length of the line, and draw arcs above and below the line.
2	What are perpendicular lines?	Lines that cross at 90°.
3	What does bisect mean?	To cut in half.
4	What is the bisector of an angle?	A line that cuts an angle in half.
5	What shape is the locus of points that are a fixed distance from a point?	A circle.
6	The locus of points a fixed distance from a line form what?	A line parallel to the original line.
7	What direction are bearings measured in?	Clockwise from north.
8	How many figures do bearings have?	Three.
9	What is a plan view?	The view from directly above.
10	What is a side elevation?	The view from the side.

Previous questions

Now go back and use these questions to check your knowledge of previous topics.

	Questions	Answers
1	What is a prism?	A 3D shape with a polygon as a constant cross-section.
2	What is the perimeter of a 2D shape?	The distance around the edges of a 2D shape.
3	What is a vertex?	A point on a 2D or 3D shape where the edges meet.
4	What is an acute angle?	An angle that measures less than 90°.
5	What is the sum of the exterior angles of any polygon?	360°

Practice 24

Exam-style questions

24.1 Here are the front and side elevations of a solid shape.

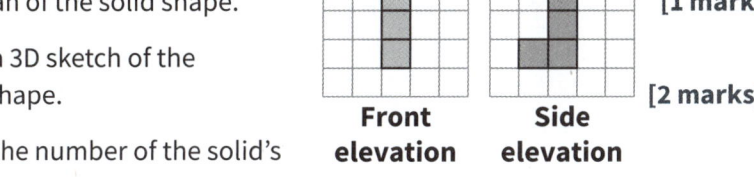

Front elevation   Side elevation

(a) Use squared grid paper to draw the plan of the solid shape. **[1 mark]**

(b) Draw a 3D sketch of the solid shape. **[2 marks]**

(c) Write the number of the solid's
  (i) vertices **[1 mark]**
  (ii) edges **[1 mark]**
  (iii) faces. **[1 mark]**

24.2 This solid object is made from six identical cubes.

Use some squared grid paper to draw the side and front elevations and the plan.

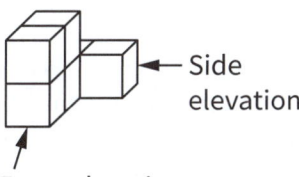

Front elevation

[3 marks]

24.3 (a) Only one of these triangles can be constructed. Which one? **[1 mark]**

  A: A triangle with sides 7 cm, 8 cm and 16 cm.
  B: A triangle with sides 3 cm, 4 cm and 6 cm.
  C: A triangle with sides 9 cm, 12 cm and 22 cm.
  D: A triangle with sides 15 cm, 30 cm and 60 cm.

(b) Using a ruler and a pair of compasses, construct the triangle you have chosen in part **a**. **[3 marks]**

Exam-style questions

24.4 Construct triangle DEF where DE = 66 mm, EF = $\frac{2}{3}$ DE and FD = $\frac{3}{4}$ EF. **[3 marks]**

24.5 Using a ruler and a pair of compasses, construct the line perpendicular to the line segment FH that passes through G. You must show your construction lines **[3 marks]**

EXAM TIP
Start by drawing arcs either side of G.

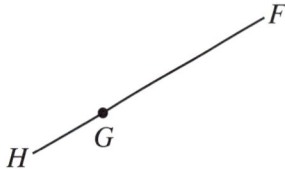

24.6 Using a ruler and a pair of compasses, constuct an angle of 45°. You must show your construction lines. **[3 marks]**

24.7 Draw a line, AB = 4.5 cm then shade the area that is further from A than B but is less than 3 cm away from B. Show all construction lines. **[4 marks]**

24.8 A triangular plot of land is shown. Milly farms the area that is closer to CD than to CE and is also less than 500 m from C.

Copy the diagram and shade the area of land that Milly farms. Show all construction lines. **[4 marks]**

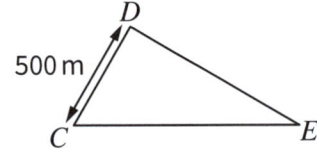

24.9 Here is a plan of a T-shaped go-kart track. There is a barrier all around the track, exactly 2.5 m from the track.
Using a scale of 1:125 draw a plan of the barrier. **[3 marks]**

24.10 A speedboat (S) is positioned due east of a dinghy (D). The bearing of a hovercraft (H) from the dinghy (D) is 080°. The bearing of the same hovercraft from the speedboat is 280°.

On the diagram, draw the position of the hovercraft. Label it H. **[3 marks]**

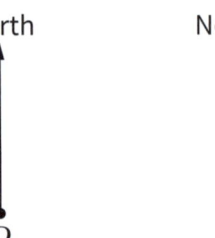

24.11 Point B is 8 km due north of point A.
Point C is 11 km due east of point A.

Work out the bearing of B from C to the nearest degree. [4 marks]

EXAM TIP
Draw a sketch.

24.12 A buoy is 30 m away from a look-out post on a bearing of 045°.
A swimmer is 40 m from the same look-out post on a bearing of 315°.

Work out the distance of the swimmer from the buoy. [4 marks]

24.13 The diagram shows the positions of 3 towns, A, B and C, on a map.
The bearing of B from A is 034°
$AB = BC$.
C is due south of B.

Work out the bearing of A from C. [3 marks]

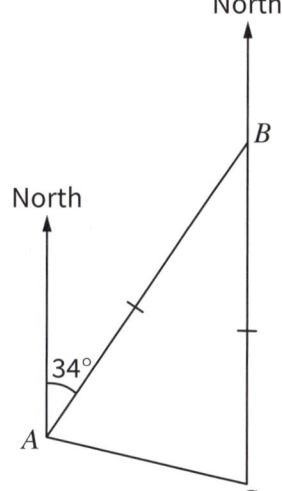

Questions referring to previous content

24.14 The length, p m, of a football pitch is given as 110 m.
Write the error interval for p if this value is rounded to

(a) the nearest 10 metres [2 marks]

(b) the nearest 5 metres. [2 marks]

24.15 A number, x, is given rounded to a particular degree of accuracy.
Write the error interval for x in each case.

(a) $x = 4.67$ to 2 decimal places [2 marks]

(b) $x = 5000$ to 1 significant figure. [2 marks]

24 Practice

Knowledge

25 Trigonometry in 3D, sine and cosine rules

Pythagoras' theorem in 3D

You can use **Pythagoras' theorem** twice to find the longest diagonal in a cuboid.

Worked example

Find the length of BG in this cuboid.

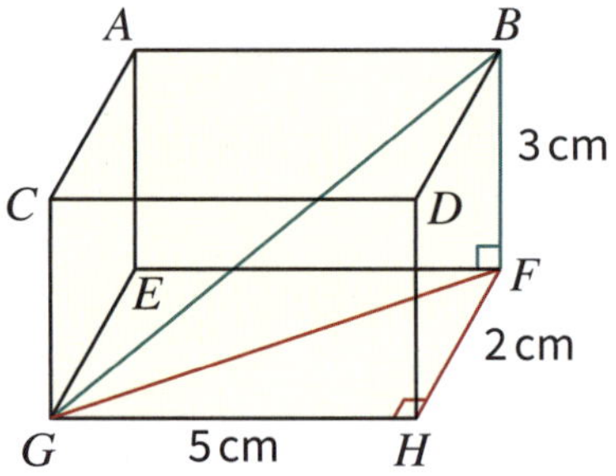

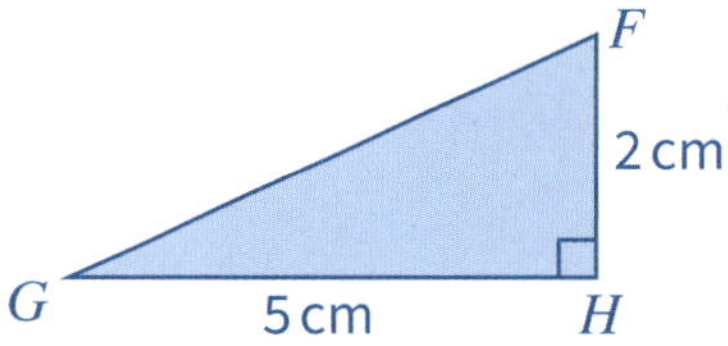

Sketch triangle FGH.

$(FG)^2 = (GH)^2 + (FH)^2$
$(FG)^2 = 5^2 + 2^2$
$FG = \sqrt{29}$ cm

Use Pythagoras' theorem to find FG.

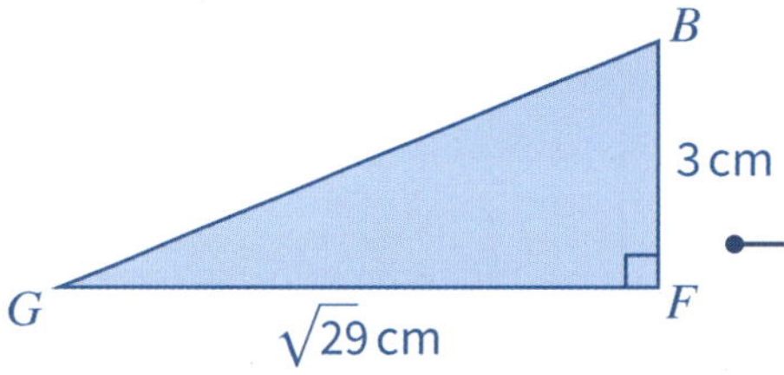

Sketch triangle BGF.

$(BG)^2 = (BF)^2 + (FG)^2$
$(BG)^2 = 3^2 + \sqrt{29}$
$BG = \sqrt{38}$ cm
$= 6.16$ cm (3 s.f.)

Use Pythagoras' theorem to find BG.

REVISION TIP

You can also use a single step to solve this problem.

$a^2 + b^2 + c^2 = d^2$

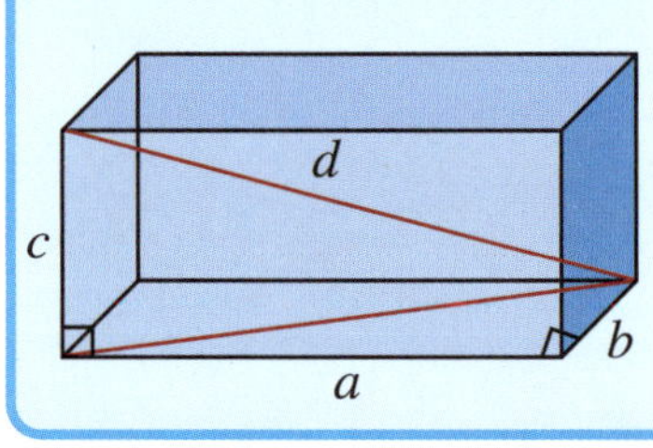

Pythagoras and trigonometry in 3D

Worked example

This pyramid has a horizontal rectangular base. Vertex A is vertically above M, which is the midpoint of EC. Calculate the size of the angle between AC and the base $BCDE$.

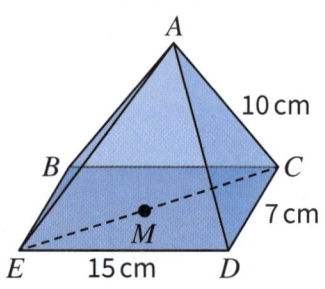

Sketch $\triangle CDE$.

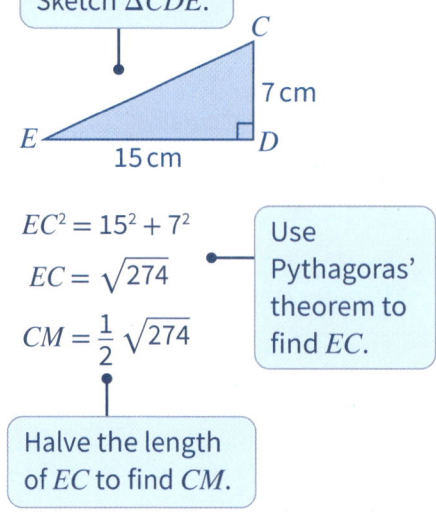

$EC^2 = 15^2 + 7^2$
$EC = \sqrt{274}$
$CM = \frac{1}{2}\sqrt{274}$

Use Pythagoras' theorem to find EC.

Halve the length of EC to find CM.

Sketch $\triangle ACM$. Use trigonometry to find angle ACM.

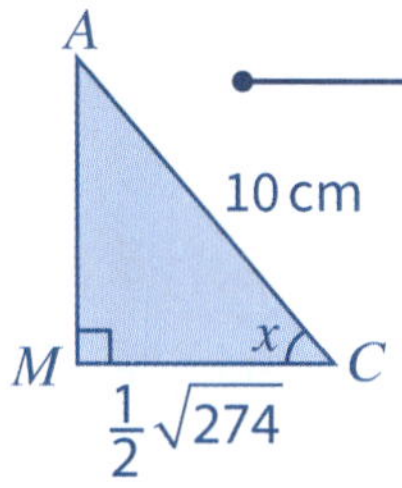

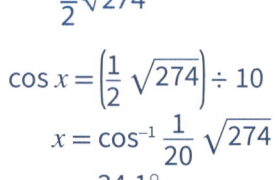

$\cos x = \left(\frac{1}{2}\sqrt{274}\right) \div 10$
$x = \cos^{-1} \frac{1}{20}\sqrt{274}$
$= 34.1°$

The angle between AC and the base is 34.1° to 1 d.p.

25

The sine and cosine rules

The **sine rule** and **cosine rule** work for all triangles, not just right-angled ones.

sine rule

To find an angle use
$$\frac{\sin A}{a} = \frac{\sin B}{b} = \frac{\sin C}{c}$$

To find a side use
$$\frac{a}{\sin A} = \frac{b}{\sin B} = \frac{c}{\sin C}$$

cosine rule

To find an angle use
$$\cos A = \frac{b^2 + c^2 - a^2}{2bc}$$

To find a side use
$$a^2 = b^2 + c^2 - 2bc \cos A$$

REVISION TIP

Opposite sides and angles have the same letter.

You can decide which angle is A, B and C.

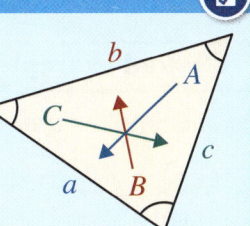

In some questions you will need to use both trigonometry of right-angled triangles, and the sine or cosine rule.

Worked example

Calculate the size of the angle marked x.

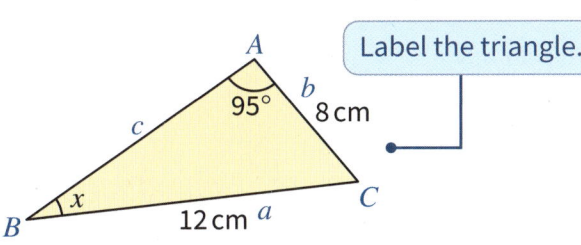

Label the triangle.

$$\frac{\sin A}{a} = \frac{\sin B}{b}$$

$$\frac{\sin x}{8} = \frac{\sin 95°}{12}$$

$$\sin x = \frac{8 \sin 95°}{12}$$

$$= 0.664...$$

$$x = \sin^{-1}(0.664...)$$

$$= 41.6° \text{ (to 1 d.p.)}$$

You know two sides and their corresponding angles, so use the sine rule.

Worked example

Calculate the length of AB.

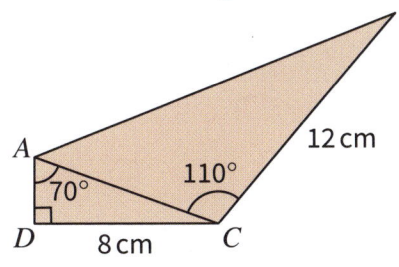

$$c^2 = a^2 + b^2 - 2ab \cos C$$

$$= 12^2 + (8.51...)^2 - 2 \times 12 \times (8.51...) \cos 110°$$

$$= 286.36...$$

$$c = \sqrt{286.36...}$$

$$= 16.9 \text{ cm (to 1 d.p.)}$$

Use the cosine rule to find side c.

Sketch triangle ADC.

$$\sin 70° = \frac{8}{AC}$$

$$AC = \frac{8}{\sin 70°}$$

$$AC = 8.51...$$

Use trigonometry to find side AC.

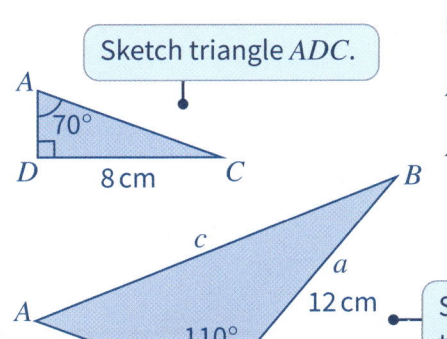

Sketch triangle ABC.

WATCH OUT

The question does not give you enough information about triangle ABC for you to use the sine or cosine rule straight away.

Knowledge

25 Trigonometry in 3D, sine and cosine rules

Area of a triangle

area of a triangle = $\frac{1}{2} ab \sin C$ where C is the angle between sides a and b.

Worked example

Calculate the area of this triangle.

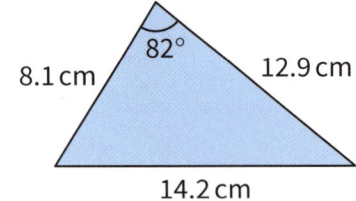

area = $\frac{1}{2} \times 8.1 \times 12.9 \times \sin 82°$
     = 51.7 cm² (1 d.p.)

Find the two sides that the marked angle is between.

Substitute these values into the equation.

Worked example

The area of triangle ABC is 128 cm².
Calculate the length of AC.

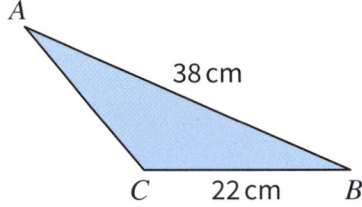

$128 = \frac{1}{2} \times 22 \times 38 \sin(\angle ABC)$

$\sin(\angle ABC) = \dfrac{128}{\frac{1}{2} \times 22 \times 38}$

$= 0.306...$

$(\angle ABC) = \sin^{-1}(0.306...)$

$= 17.8°$

Substitute the known values into the equation for area of a triangle using the sine rule.

Using the cosine rule gives

$AC^2 = 22^2 + 38^2 - 2 \times 22 \times 38 \times \cos(17.8...°)$

$= 336.3$

$AC = \sqrt{336.3...}$

$= 18.3$ cm (1 d.p.)

Substitute the known values into the cosine rule equation to find a side.

Key terms Make sure you can write a definition for these key terms

cosine rule   Pythagoras' theorem   sine rule

25 Trigonometry in 3D, sine and cosine rules

Learn the answers to the questions below, then cover the answers column with a piece of paper and write as many as you can. Check and repeat.

Questions / Answers

1. State Pythagoras' theorem. — $a^2 + b^2 = c^2$

2. What is the formula $a^2 + b^2 + c^2 = d^2$ used for? — To find the longest diagonal in a cuboid.

3. In the previous formula, what are a, b, and c? — The sides of the cuboid.

4. What is the sine rule to find an angle? — $\dfrac{\sin A}{a} = \dfrac{\sin B}{b} = \dfrac{\sin C}{c}$

5. What is the sine rule to find a length? — $\dfrac{a}{\sin A} = \dfrac{b}{\sin B} = \dfrac{c}{\sin C}$

6. What is the cosine rule to find a side? — $a^2 = b^2 + c^2 - 2bc \cos A$

7. What is the cosine rule to find an angle? — $\cos A = \dfrac{b^2 + c^2 - a^2}{2bc}$

8. How do you label a related side and an angle? — A related side and angle are opposite each other.

9. What is the formula to find the area of a triangle using trigonometry? — area $= \dfrac{1}{2} ab \sin C$

10. The sine rule states that $\dfrac{\sin A}{a} = \dfrac{\sin B}{b}$. Rearrange the equation to make A the subject. — $A = \sin^{-1} \dfrac{a}{b} \sin B$

Previous questions

Now go back and use these questions to check your knowledge of previous topics.

Questions / Answers

1. How do we know when a ratio is written in its simplest form? — The only remaining common factor of each of the parts of the ratio is 1

2. What is the general form of a quadratic equation? — $y = ax^2 + bx + c$

3. Give the exact value of tan 30° — $\dfrac{1}{\sqrt{3}}$ (which can also be written as $\dfrac{\sqrt{3}}{3}$)

4. What is the term-to-term rule? — The rule that gets you from one term to the next, which is the same for every term.

5. How do you solve a quadratic equation that can be factorised? — Rearrange the equation such that $ax^2 + bx + c = 0$, then factorise.

Practice

Exam-style questions

25.1 The diagonal of the cuboid shown is 13 cm. The length of the cuboid is 12 cm and its width is 3 cm. Work out the height, h, of the cuboid. **[3 marks]**

EXAM TIP
Use Pythagoras' theorem to write an equation for the diagonal.

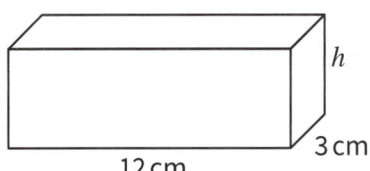

25.2 The 3D shape shown is made using 1 cm cubes. Vertices P and Q are labelled. Work out the length of the line PQ. **[2 marks]**

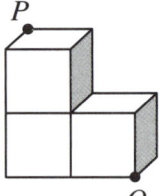

25.3 $ABCDE$ is a square-based pyramid with vertex E directly above the centre of the base of the pyramid. The square base has a side length $\sqrt{2}$ cm and AE is $\sqrt{3}$ cm. Calculate the height of the pyramid. **[3 marks]**

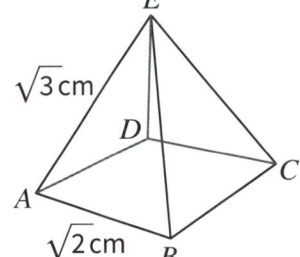

25.4 $VWXY$ is a tetrahedron with faces that are all identical equilateral triangles of side length 3 m. The vertex Y is directly above the centre of the base of the tetrahedron.

Work out the angle between the side XY and the base of the tetrahedron to the nearest degree. **[4 marks]**

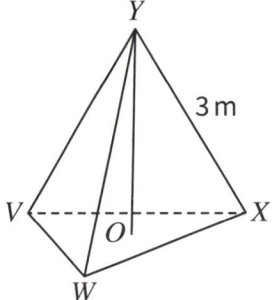

25.5 The diagram shows a triangular prism. The base, $PQRS$, of the prism is a rectangle with length 20 cm and width 12 cm. Angle $POS = 38°$. T is the point on SR such that $ST:TR = 7:3$.

Calculate the angle between OT and the base of the prism. Give your answer correct to 3 significant figures. **[5 marks]**

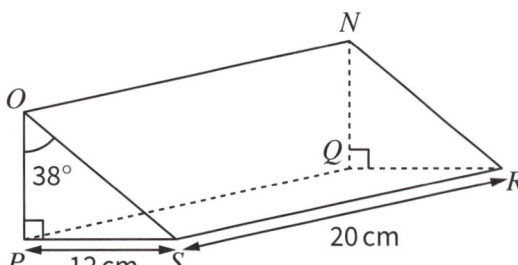

25.6 The diagram shows a cuboid. O is a point at the centre of the base of the cuboid. FC = 5 cm, AB = 4 cm and HE = 6 cm.

Work out the angle between FO and the base of the cuboid. [5 marks]

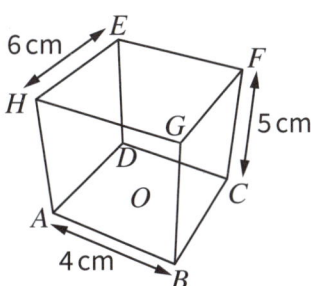

25.7 Work out the length x in each triangle shown. Give your answers correct to 1 decimal place. [6 marks]

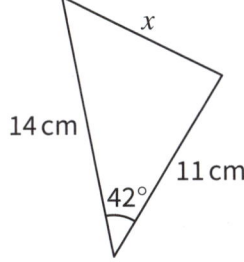 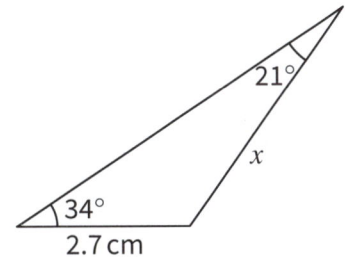

25.8 Work out the angle x in each triangle shown. Give your answers correct to 1 decimal place. [6 marks]

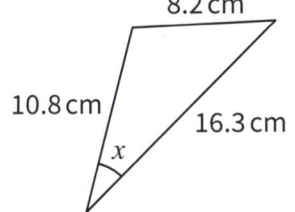 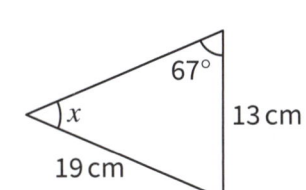

25.9 A triangle has vertices A, B and C. AB = 10 cm, BC = 6 cm and angle CAB is 25°.

Given that angle ACB is obtuse, work out angle ACB to 1 decimal place. [4 marks]

EXAM TIP
Draw a sketch.

25.10 Work out the length of side BC in the diagram. Give your answer to 1 decimal place. [5 marks]

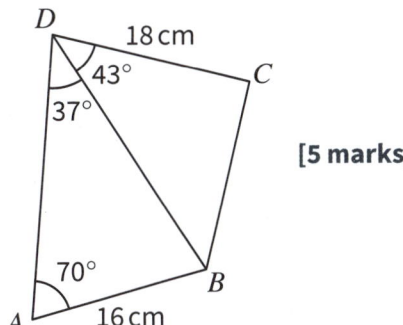

Exam-style questions

25.11 Work out the area of each triangle shown.
Give your answers correct to 2 decimal places. **[4 marks]**

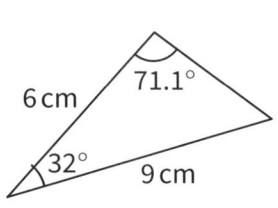
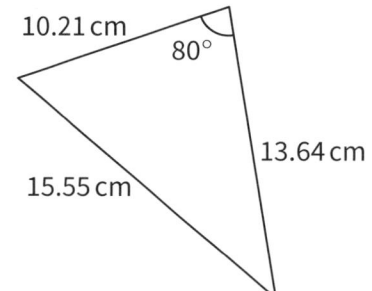

25.12 A triangular lawn has sides of length 13 m, 11 m and 4 m.

Work out the area of the lawn correct to 3 significant figures. **[4 marks]**

25.13 A triangular-based pyramid consists of four equilateral triangles of side length 8 cm. Work out the exact surface area of the pyramid. State your units. **[4 marks]**

25.14 Work out the area of quadrilateral $ABCD$. Give your answer correct to 1 decimal place. **[5 marks]**

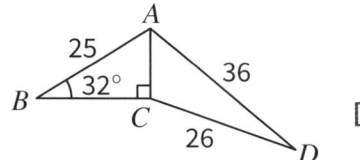

25.15 A cylindrical container is three-quarters full of water. The diameter of the container is 11 cm. A solid triangular prism is added to the container and is completely submerged, with the water level rising to the very top of the container without overflowing. The cross-section of the prism is an equilateral triangle of side length 4 cm. The diagonal of each rectangular side of the prism is 14 cm.

Work out the height of the container to 3 significant figures. **[5 marks]**

25.16 The diagram shows the intersection of a circle, centre O, and an equilateral triangle, OAB. $AB = 8$ cm. C and D are the midpoints of OB and OA respectively.

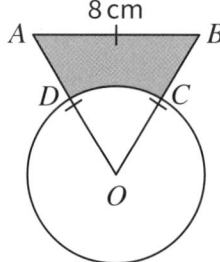

(a) Work out the exact area of the shaded region $ABCD$. **[4 marks]**

(b) A chord is drawn on the circle from C to D. Work out, to 1 decimal place, the area of the minor segment formed. **[3 marks]**

25.17 ABC and BCD are triangles.
The area of triangle ABC is 80 cm².

Work out the size of angle CBD.
Give your answer to 3 significant figures.

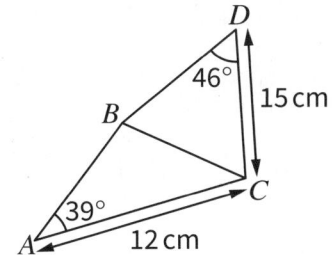

[5 marks]

Questions referring to previous content

25.18 Rick writes a 2-digit odd number.

(a) Work out how many different numbers Rick could write.

Amir writes a 2-digit number. [2 marks]

(b) What is the probability that Amir writes a number whose digits are both the same? [2 marks]

 25.19 Triangle ABC has vertices $A(0, 0)$, $B(2, -2)$ and $C(-1, -4)$

Find the vertices of A'B'C' when ABC is reflected in the line $x = 1$ [2 marks]

Knowledge

26 Circle theorems and circle geometry

Circle theorems

Angles in the same **segment** are equal.

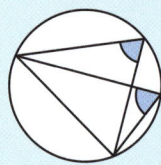

An angle in a **semicircle** is a right angle.

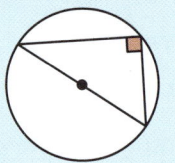

LINK

To remind yourself about the parts of a circle, look back at Chapter 19.

An angle at the centre is twice the angle at the **circumference**.

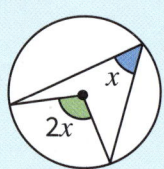

Opposite angles in a **cyclic quadrilateral** add up to 180°.

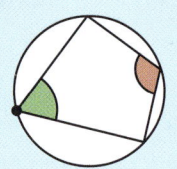

REVISION TIP

All the vertices of a cyclic quadrilateral lie on the circumference of a circle.

If you are asked to prove a **circle theorem**, use angle facts and other circle theorems to clearly justify your working at each stage.

Worked example

A, B and C are points on the circumference of a circle.
The centre of the circle is O.

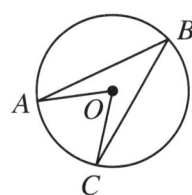

Prove that $\angle AOC = 2 \angle ABC$

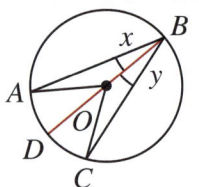

$AO = OB$, since both AO and OB are radii. — State the facts that apply to $\triangle ABO$.
Hence, $\triangle ABO$ is isosceles.

If $\angle OBA = x$, then $\angle OAB = x$, since base angles of an isosceles triangle are equal. — Label $\angle OBA$ as x. Consider $\triangle ABO$.

$\angle AOB = 180 - 2x$, since angles in $\triangle ABO$ sum to 180°.

$\angle DOA = 180 - \angle AOB$ — Use triangle facts to deduce $\angle DOA$.
$\qquad = 180 - (180 - 2x)$
$\qquad = 2x$

$\angle DOC = 2y$ (as above) — Use triangle facts to deduce $\angle AOC$.

$\angle ABC = x + y$

$\angle AOC = 2x + 2y$
$\qquad = 2(x + y)$
$\qquad = 2\angle ABC$

— Split $ABCO$ into $\triangle ABO$ and $\triangle CBO$.

Circle theorems

> **Worked example**
>
> Given that $\angle ABC = 35°$, calculate the size of $\angle CAO$.
>
>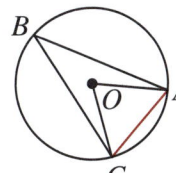
>
> $\angle AOC = 2 \times \angle ABC$ — Angle at the centre is twice the angle at the **circumference**.
> $= 2 \times 35°$
> $= 70°$
>
> In $\triangle CAO$, $AO = AC$ — AO and AC are radii and all radii in a circle must be equal.
> Hence $\triangle CAO$ is isosceles.
>
> $\angle CAO = \dfrac{180 - 70}{2}$ — The value of $\angle AOC$ is $70°$ and two angles of an isosceles triangle must be equal.
> $= 55°$

The angle between a tangent and the radius is 90°.

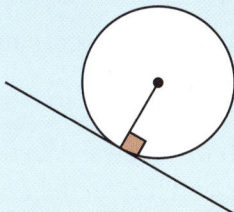

Two tangents from the same point are equal in length.

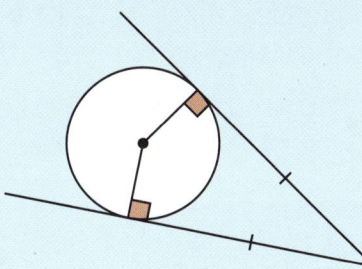

The angle between a tangent and a **chord** is equal to the angle from the chord in the **alternate segment** of the circle.

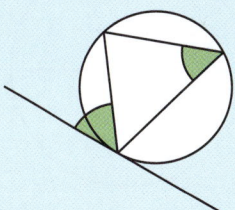

> **Worked example**
>
> Work out the size of angles x and y.
> State any circle theorems that you use.
>
>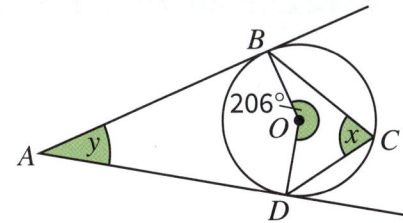
>
> obtuse $\angle BOD = 360 - 206$
> $= 154°$
> $x = \dfrac{154}{2} = 77°$
>
> (angle at centre = angle at circumference $\times$ 2)
>
> $\angle ABO = \angle ADO = 90°$ (angle between tangent and radius is 90°)
>
> $y = 360 - 90 - 90 - 154$
> $= 26°$ (angles in a quadrilateral add up to 360°)

26 Knowledge

Knowledge

26 Circle theorems and circle geometry

Circle geometry

The equation of a circle with centre (0, 0) and radius r is

$$x^2 + y^2 = r^2$$

This comes from Pythagoras' theorem.

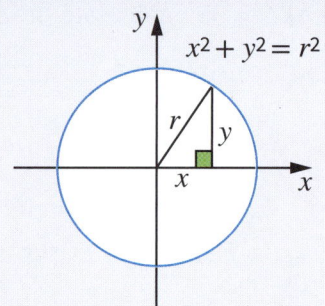

Worked example

1. Write the equation of the circle shown.

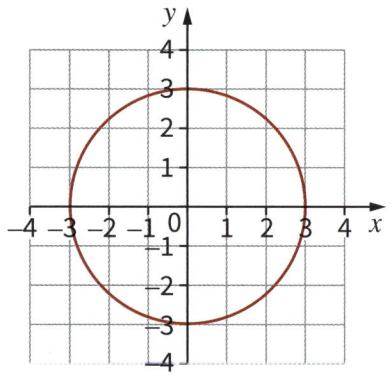

The centre of the circle is the origin and its radius is 3, so its equation is

$x^2 + y^2 = 9$

Substitute 3 in place of r

2. Show that the point $(-1, 2\sqrt{2})$ lies on the circle.

$x^2 + y^2 = (-1)^2 + (2\sqrt{2})^2$
$= 1 + 8$
$= 9$

so $(-1, 2\sqrt{2})$ lies on the circle.

Substitute the coordinates into the equation

If the answer is 9, the point lies on the circle.

LINK

To remind yourself about multiplying surds, look back at Chapter 3.

Equation of the tangent to a circle at a point on the circumference

The equation of the tangent to a circle at a point on the circumference is $y = mx + c$, where:
m = the gradient of the line c = the intercept of the line at the y-axis

Worked example

A circle with centre (0, 0) has the equation $x^2 + y^2 = 100$

The point A (6, −8) lies on the circle.
Work out the equation of the tangent to the circle at A.

gradient of $OA = -\frac{8}{6} = -\frac{4}{3}$ — Find the gradient of the radius OA.

gradient of tangent at

$A = \frac{3}{4}$

$y = mx + c$

$-8 = \frac{3}{4} \times 6 + c$

$c = -\frac{25}{2}$

$y = \frac{3}{4}x - \frac{25}{2}$

Use $m_1 m_2 = -1$ for perpendicular lines to find gradient of tangent at A.

Substitute $m = \frac{3}{4}$ and $(x, y) = (6, -8)$ into $y = mx + c$ to find c.

Write the equation of the tangent.

26

Equation of the tangent to a circle at a point on the circumference

Worked example

Solve these simultaneous equations.
Show your working. — Number the equations.

$2x + y = 10$ ①

$x^2 + y^2 = 40$ ②

$y = 10 - 2x$ ③ — Rearrange ① to get y on its own. Label this ③.

$x^2 + (10 - 2x)^2 = 40$ — Substitute y into ②.

$x^2 + 100 - 40x + 4x^2 = 40$

$5x^2 - 40x + 60 = 0$

$x^2 - 8x + 12 = 0$ — Solve for x.

$(x - 2)(x - 6) = 0$

$x = 2$ or $x = 6$

When $x = 2$, — Substitute the two values of x into ③ to find y.

$\quad y = 10 - 2(2)$

$\quad\quad = 6$

When $x = 6$,

$\quad y = 10 - 2(6)$

$\quad\quad = -2$

REVISION TIP

This solution tells you that the graphs of $2x + y = 10$ and $x^2 + y^2 = 40$ intersect at the points (2, 6) and (6, −2).

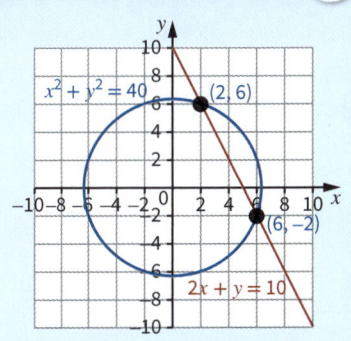

WATCH OUT

Double check your solutions by substituting them into one of the other equations and checking that it balances.

Worked example

The graph of $x^2 + y^2 = 4$ is shown.

Use a graphical method to find an estimate for the solutions to the simultaneous equations:

$x^2 + y^2 = 4$ ①

$y = 1 - \frac{1}{2}x$ ② — Number the equations.

$x = 2, y = 0$ and $x = -1.2, y = 1.6$ — Estimate the coordinates for the two points where the graphs intersect.

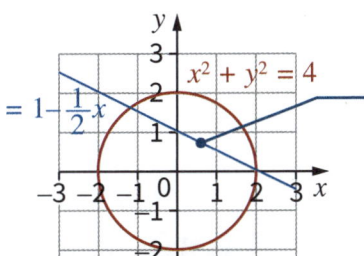

Draw the graph of $y = 1 - \frac{1}{2}x$ on the same axes.

Key terms — Make sure you can write a definition for these key terms

alternate segment chord circle theorem
circumference cyclic quadrilateral segment semicircle

26 Knowledge

Retrieval

26 Circle theorems and circle geometry

Learn the answers to the questions below, then cover the answers column with a piece of paper and write as many as you can. Check and repeat.

Retrieval | Answers

1. Write the relevant theorem/fact for this diagram. — Angles in the same segment are equal.

2. Opposite angles in a cyclic quadrilateral add up to what? — 180°

3. An angle at the centre is twice the angle of what? — The circumference.

4. What is the angle in a semicircle? — A right angle.

5. Two tangents from the same point are equal in what? — Length.

6. What is the angle between a tangent and the radius? — 90°

7. The angle between a tangent and a chord is equal to what? — The angle from the chord in the alternate segment of the circle.

8. What is the general equation of a circle that has its centre on the origin? — $x^2 + y^2 = r^2$

Previous questions
Now go back and use these questions to check your knowledge of previous topics.

Questions | Answers

1. What does $2a$ mean? — 2 multiplied by a

2. How can you check your solutions to a pair of simultaneous equations? — Substitute your values back into the equations.

3. What is the area of a 2D shape? — The space inside a 2D shape.

4. How do you find the volume of a prism? — Multiply the area of the cross-section by the length.

5. What units do you use for volume? — Cubic units.

Practice 26

Exam-style questions

26.1 The circle $x^2 + y^2 = 1$ has centre at (0, 0) and radius 1.
Lin says that the circle must pass through the point (1, 1).

Show that Lin is wrong. **[2 marks]**

26.2 The graph shows a circle, centre (0, 0).

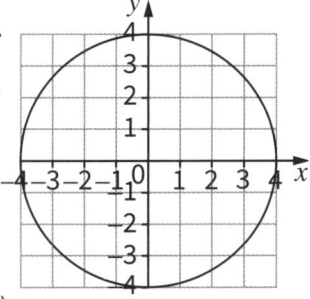

(a) Write the equation of the circle. **[1 mark]**

(b) Show that the point $(2\sqrt{2}, 2\sqrt{2})$ lies on the circle. **[2 marks]**

(c) Write the equation of the tangent to the circle that passes through the point (0, −4). **[1 mark]**

26.3 The circle, C, has equation $x^2 + y^2 = 5$. Find the equation of the tangent to C at the point (1, 2). **[5 marks]**

EXAM TIP
Always draw a sketch if you're not given one.

26.4 A graph has the equation $x^2 + y^2 = 10$. It passes through the points A, B, C and D, which lie on the axes.

Prove that $ABCD$ is a square. **[4 marks]**

26.5 Write the size of angle x in each of these circles, giving a reason for your answer. **[6 marks]**

(a) 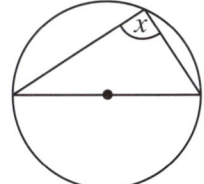 (b) 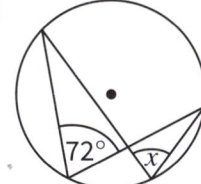 (c)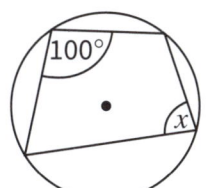

26.6 (a) Write the size of angle y in the diagram, stating the circle theorem used. **[2 marks]**

(b) Prove that angles in the same segment of a circle are equal. You may use, without proof, the circle theorem you used in part **a**. **[4 marks]**

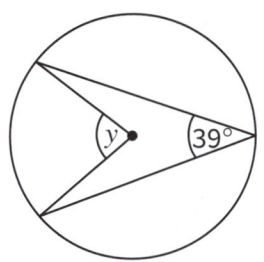

Exam-style questions

26.7 Work out the size of angle z. You must give a reason for each stage of your working. **[5 marks]**

> **EXAM TIP**
> State the names of any circle theorems used and show all calculations clearly.

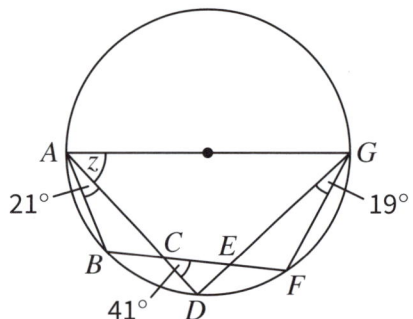

26.8 Write the value of x in each of these diagrams, giving a reason for your answer. **[6 marks]**

(a)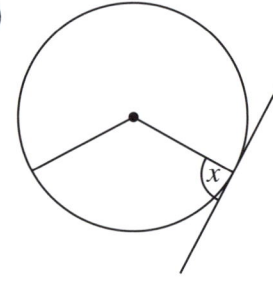

(b)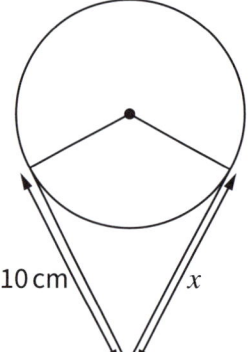

(c)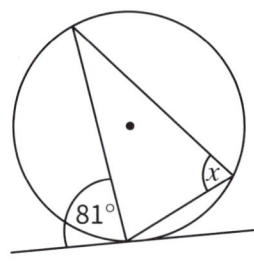

26.9 In the diagram, lines AB and AC are both tangents to the circle. $BD = CD$. Work out the size of angle p. You must give a reason for each stage of your working. **[4 marks]**

> **EXAM TIP**
> State the names of any circle theorems used and show all calculations clearly.

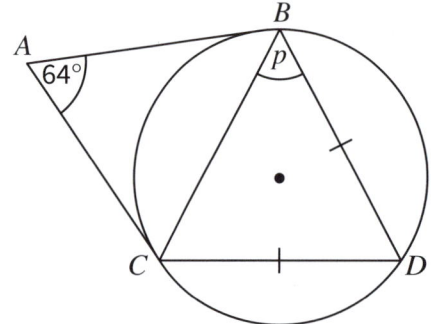

26.10 Prove that two tangents to a circle, which meet at a point outside of the circle, are equal in length. You may use, without proof, the fact that the angle between the tangent and radius of a circle is a right angle. **[4 marks]**

26.11 A circle has equation $x^2 + y^2 = 25$.

(a) Show that point A with coordinates $(-4, 3)$ lies on this circle. **[1 mark]**

(b) On a squared grid, make a sketch of the circle, indicating the coordinates of the centre and the coordinates where the circle intersects the axes. **[2 marks]**

(c) A right-angled triangle, ABC, has each of its vertices on the circle. The coordinates of A and B are (−4, 3) and (3, 4) respectively.

Given that vertex C has a negative y-coordinate, find the two possible positions of vertex C. Give a reason for your answer. **[3 marks]**

26.12 Prove algebraically that the straight line with equation
$x + y + 6 = 0$ is a tangent to the circle $x^2 + y^2 = 18$. **[4 marks]**

26.13 OAC is a sector of a circle, centre O. AB is a tangent to the circle at point A. CB is a tangent to the circle at point C. AB = 8 cm. Angle AOC = 100°.

Calculate the area of the shaded region. Give your answer to 3 significant figures. **[4 marks]**

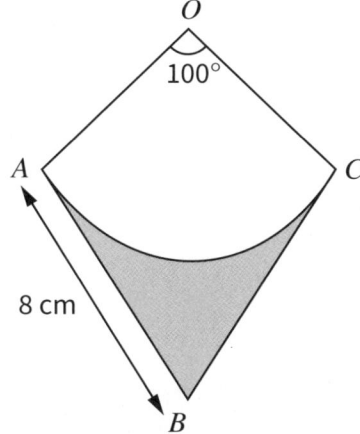

26.14 V, W and X are points of the circumference of a circle, centre O. YXZ is the tangent to the circle at the point X. Angle VXY = 84°. Angle OXW = 39°.

Work out the size of angle OVW. You must show your working. **[4 marks]**

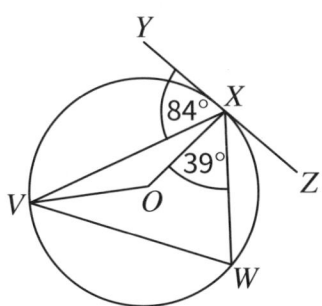

Questions referring to previous content

26.15 The first three terms of a geometric sequence are $x + 2$, 6, $9x + 3$, where $x > 0$

Find the exact value of the 5th term of the sequence. **[5 marks]**

26.16 Triangle ABC has sides of lengths AB = 8.8, BC = 7.8 and AC = 5.6.

Find the angle at vertex A to the nearest tenth of a degree. **[3 marks]**

Knowledge

27 Vectors

Vector notation

A **vector** is a way of describing movement.
Vectors have both **direction** and **magnitude** (size).

One way to write a vector is as a **column vector**.

Other notations for representing vectors include:

An arrow	A bold letter	An underlined letter
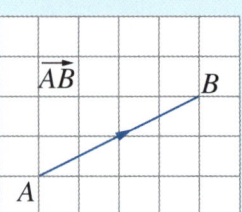	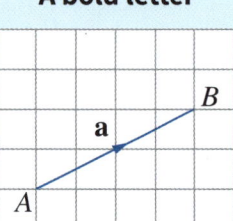	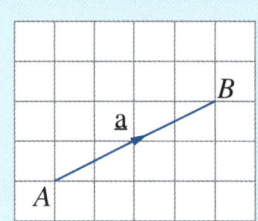

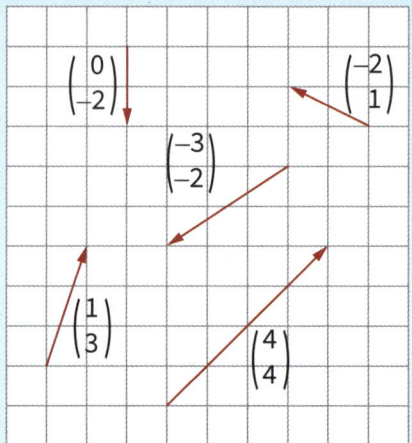

In the example above, the vector $\overrightarrow{BA}$ could be written as $-\mathbf{a}$ or $-\underline{a}$.
It is a parallel vector of the same length but opposite direction.

LINK

You revised translations of shapes using column vectors in Chapter 23. The top number tells you how far right or left the shape moves, and the bottom number tells you how far up or down it moves.

Multiplying a vector by a scalar

A **scalar** is a quantity with just size. A number is a scalar.

Scalars can be used as scale factors or multipliers.

To multiply column vectors by a scalar, multiply the top and bottom values by the number.

For example:

$3\begin{pmatrix}-1\\8\end{pmatrix}=\begin{pmatrix}3\times-1\\3\times8\end{pmatrix}=\begin{pmatrix}-3\\24\end{pmatrix}$

Worked example

$\mathbf{a}=\begin{pmatrix}2\\3\end{pmatrix}$. Using this information, draw $\mathbf{a}$ and $2\mathbf{a}$ on a grid.

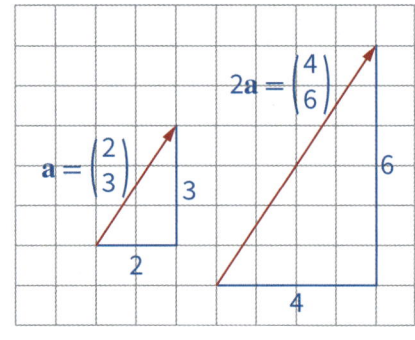

2 is a scalar. It only has magnitude (size).

$2\mathbf{a}=2\begin{pmatrix}2\\3\end{pmatrix}=\begin{pmatrix}2\times2\\2\times3\end{pmatrix}=\begin{pmatrix}4\\6\end{pmatrix}$

a is a vector. It has both direction and magnitude.

Notice how **a** and **2a** are parallel.

27

Parallel vectors

On this diagram, vector $\overrightarrow{CD}$ is also **a** as it has the same size and direction as the vector $\overrightarrow{AB}$.

Two vectors are **parallel** if one is a multiple of another.

The vector −**a** is the same size as **a** and parallel to **a** because it is $-1 \times$ **a**, but it is in the opposite direction to **a**.

A vector that is parallel to **a** but a different length will be a multiple of **a**, such as 2**a**.

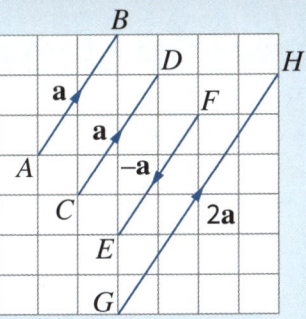

Worked example

$OABC$ is a parallelogram.

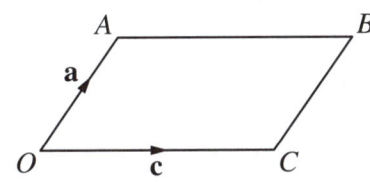

$\overrightarrow{OA} = $ **a** and $\overrightarrow{OC} = $ **c**

Write down these vectors in terms of **a** and **c**

$\overrightarrow{CO} = -$**c** — it is parallel to $\overrightarrow{OC}$ but in the opposite direction.

$\overrightarrow{OB} = \overrightarrow{OA} + \overrightarrow{AB} = $ **a** + **c**

$\overrightarrow{AC} = \overrightarrow{AO} + \overrightarrow{OC} = -$**a** + **c** — you could write **c** − **a**, the order doesn't matter.

REVISION TIP: If two vectors are parallel, one is a multiple of the other.

REVISION TIP: Remember to underline the letter if you write vectors by hand.

If M is the midpoint of AB, then $\overrightarrow{AM} = \frac{1}{2}\overrightarrow{AB}$.

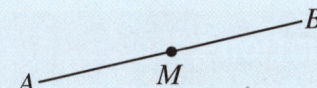

If two line segments are parallel and both pass through a common point, they must lie on a straight line.

To show that points A, B and C lie on a straight line, you need to show that any two of $\overrightarrow{AB}$, $\overrightarrow{AC}$ or $\overrightarrow{BC}$ are multiples of each other.

Worked example

$ABCD$ is a trapezium. $\overrightarrow{AB} = 2\overrightarrow{DC}$ and $\overrightarrow{AF} = 2\overrightarrow{AB}$.

E is a point on BC such that $\overrightarrow{CE} : \overrightarrow{EB} = 1 : 2$

Show that DEF is a straight line.

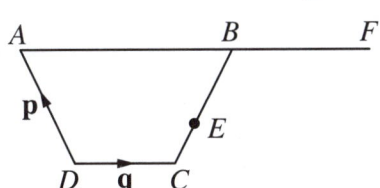

$\overrightarrow{DE} = \overrightarrow{DC} + \frac{1}{3}\overrightarrow{CB}$ since $CE : EB = 1 : 2$

$= q + \frac{1}{3}(-q + p + 2q)$ — $\overrightarrow{AB} = 2\overrightarrow{DC} = 2q$

$= \frac{1}{3}p + \frac{4}{3}q$

$\overrightarrow{DF} = \overrightarrow{DA} + 2\overrightarrow{AB}$ — $\overrightarrow{AF} = 2\overrightarrow{AB}$

$= p + 2(2q)$

$= p + 4q$

$= 3\overrightarrow{DE}$

Knowledge

27 Vectors

Adding and subtracting vectors

To add column vectors:
- add the top values together
- add the bottom values together.

Worked example

Given that $\mathbf{a} = \begin{pmatrix} 4 \\ -3 \end{pmatrix}$ and $\mathbf{b} = \begin{pmatrix} 1 \\ 7 \end{pmatrix}$

1. calculate $\mathbf{a} + \mathbf{b}$

$\begin{pmatrix} 4 \\ -3 \end{pmatrix} + \begin{pmatrix} 1 \\ 7 \end{pmatrix}$

$\begin{pmatrix} 4+1 \\ -3+7 \end{pmatrix} = \begin{pmatrix} 5 \\ 4 \end{pmatrix}$

Choose a starting point and draw vector **a**

2. show the vectors **a**, **b** and **a** + **b** on a squared grid.

Starting from the end point of the first vector, draw vector **b**

The sum of the two vectors is the single vector that takes you from the starting point of the first vector to the end point of the second vector.

To subtract column vectors:
- subtract the top value of the second vector from the top value of the first
- subtract the bottom value of the second vector from the bottom value of the first.

Using simultaneous equations

You can form a pair of simultaneous equations to solve for x and y.

Worked example

$4\begin{pmatrix} x \\ 1 \end{pmatrix} = \begin{pmatrix} -y \\ 2x \end{pmatrix}$ Work out the values of x and y.

$4x = -y$ ①

$4 = 2x$ ②

$\frac{4}{2} = \frac{2x}{2} \Rightarrow x = 2$

$4 \times 2 = -y \Rightarrow y = -8$

$x = 2, y = -8$

Equate the top components and bottom components, and write as a pair of simultaneous equations.

From ②

Substitute $x = 2$ in ①

Key terms — Make sure you can write a definition for these key terms

column vector   direction   magnitude
parallel   scale factor   vector

Retrieval 27

27 Vectors

Learn the answers to the questions below, then cover the answers column with a piece of paper and write as many as you can. Check and repeat.

Questions | Answers

1. What does magnitude mean? — Magnitude means size.
2. What is a scalar? — A quantity that has only size.
3. What are the different ways that this vector can be written? 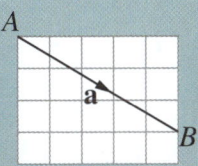 — $\vec{AB}$ or **a** or <u>a</u> or $\begin{pmatrix} 5 \\ -3 \end{pmatrix}$
4. How do you know when vectors are parallel? — They are multiples of each other.
5. What do you know about the relationship between **a** and **−a**? — **−a** is the same size as **a** and is parallel to **a**, but is in the opposite direction.
6. What do you know about the relationship between **a** and **2a**? — **2a** is parallel to **a**, and in the same direction, but it is twice as long.
7. What does the top value in a column vector tell you? — How far to move to the right, if the number is positive and how far to move to the left, if the number is negative.
8. How do you add column vectors? — Add the top values together and add the bottom values together.
9. How do you multiply a column vector by a scalar? — Multiply both the top and bottom values by the number.
10. How can you represent the difference of two vectors, **a − b** on a grid? — **−b** is the same as **b**, but in the opposite direction. **a − b** is drawn as the sum of **a** and **−b**.
11. How can you show that points A, B and C lie on a straight line? — Show that any two of $\vec{AB}$, $\vec{BC}$ and $\vec{AC}$ are parallel.

Previous questions

Now go back and use these questions to check your knowledge of previous topics.

Questions | Answers

1. What are the four types of transformations? — Reflections, rotations, translations, enlargements.
2. What is a mixed number? — A number that has a whole number and a fraction.
3. What is an inequality? — A relationship between two expressions or values that are not equal to each other.
4. What are the two algebraic methods you can use to solve simultaneous equations? — Elimination and substitution.

Practice

Exam-style questions

27.1 The vector **a** and the vector **b** are shown on the grid.

(a) On the grid, draw and label vector **a** − **b**. [2 marks]

(b) Work out 2**b** + 3**a** as a column vector. [2 marks]

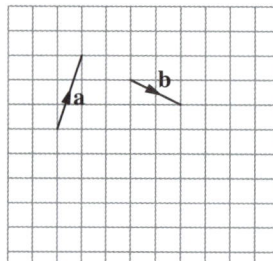

27.2 PQRSTU is a regular hexagon. Write these vectors in terms of **a** and **b**.

(a) $\overrightarrow{OT}$ (b) $\overrightarrow{PQ}$

(c) $\overrightarrow{OU}$ (d) $\overrightarrow{UQ}$

[4 marks]

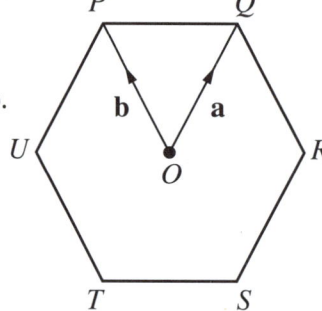

27.3 **p** and **q** are vectors such that $\mathbf{p} = \begin{pmatrix} 4 \\ -1 \end{pmatrix}$ and $2\mathbf{q} - 5\mathbf{p} = \begin{pmatrix} -26 \\ 15 \end{pmatrix}$.

Find **q** as a column vector. [3 marks]

27.4 In the grid shown, $\overrightarrow{OA} = \begin{pmatrix} 0 \\ 2 \end{pmatrix}$, $\overrightarrow{OB} = \begin{pmatrix} 2 \\ 4 \end{pmatrix}$ and $\overrightarrow{OD} = \begin{pmatrix} 1 \\ 1 \end{pmatrix}$.

Given that ABCD is a rectangle, write the column vector that represents $\overrightarrow{OC}$. [2 marks]

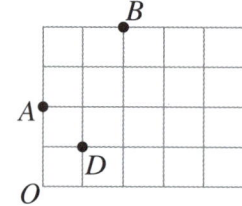

27.5 In the diagram, $\overrightarrow{OA} = \mathbf{a}$ and $\overrightarrow{OB} = \mathbf{b}$.

M is the midpoint of OA and N lies on AB such that AN : NB = 4 : 1.

Find $\overrightarrow{MN}$ in terms of **a** and **b**. [4 marks]

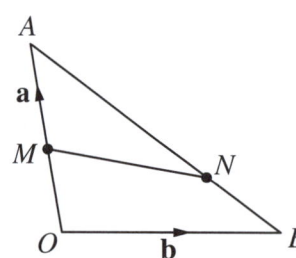

27.6 In the diagram, $\overrightarrow{OP} = \mathbf{p}$ and $\overrightarrow{OQ} = \mathbf{q}$.

PQR is a straight line such that PQ : QR = 2 : 3.

Show that $\overrightarrow{OR}$ is parallel to 5**q** − 3**p**. [5 marks]

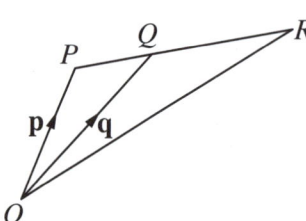

EXAM TIP

Parallel vectors are a constant multiple of each other.

27.7 OAB is a triangle. OPA and OQB are straight lines.

$\overrightarrow{OA} = \mathbf{a}$ and $\overrightarrow{OB} = \mathbf{b}$. M is the midpoint of QA and lies on the line PB. OP : PA = 3 : 1.

Work out the ratio OQ : QB. [5 marks]

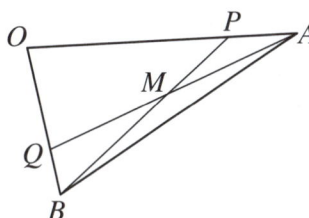

27.8 OABC is a trapezium with OA parallel to CB.

OA = 3CB.

$\overrightarrow{OA} = 9\mathbf{a}$ and $\overrightarrow{OC} = \mathbf{c}$.

Point P lies on AC such that AP : PC = 3 : 1.

Prove that O, P and B are collinear. [4 marks]

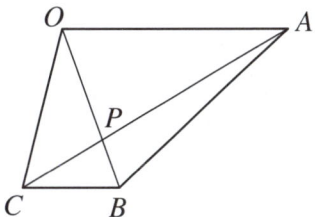

Questions referring to previous content

27.9 At time $t = 0$, Riley passes point A, running at a constant speed of 8 km/h. At the same instant, Jordan, running in the opposite direction, passes point B at a constant speed of 2.5 m/s. The distance between A and B is 8500 m.

(a) Which of the runners is travelling faster? Show your working. [2 marks]

(b) Work out the time in minutes when the runners pass each other. [3 marks]

(c) Work out how far the runners will be from A when they pass each other. [2 marks]

27.10 Prove that the angle in a semicircle is a right angle.

Do not use circle theorems in your proof.

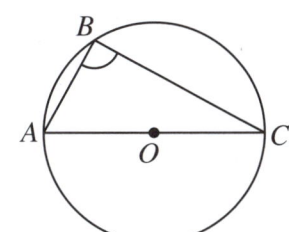

[4 marks]

Knowledge

28 Probability

Probability basics

Probability is the likelihood that an event will happen.

It is always a value between 0 and 1.

Probabilities can be written as fractions, decimals or percentages.

The probability scale

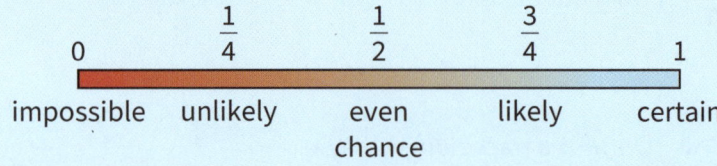

Theoretical probability

A **theoretical probability**, written as P(event), is the probability that an event will happen.

If all possible outcomes are equally likely, then the theoretical probability is:

$$P(\text{event}) = \frac{\text{number of ways event can happen}}{\text{total number of possible outcomes}}$$

For example, the probability of rolling a six with a dice is

$P(\text{six}) = \frac{1}{6}$ — There is one six on a dice.
— There are six numbers on a dice.

Mutually exclusive events are events that can't happen at the same time.

For example, a number cannot be both odd and even.

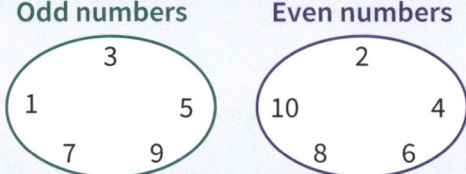

The probabilities of all possible mutually exclusive outcomes must add up to 1.

P(event happening) = 1 − P(event not happening)

Worked example

Each card in a pack of cards shows a shape that may be a triangle, a square, a pentagon, or a hexagon.

The table shows the probabilities of selecting a triangle or a square.

The ratio of pentagons to hexagons in the pack is 5 : 7

Complete the table.

Shape	Triangle	Square	Pentagon	Hexagon
Probability	0.38	0.26	0.15	0.21

$1 − (0.38 + 0.26) = 0.36$ — The four probabilities must add to 1.

$P(\text{pentagon}) : P(\text{hexagon}) = 5 : 7$

$P(\text{pentagon}) = \frac{5}{12} \times 0.36$ — Use the ratio of pentagons to hexagons to write the fraction for the number of pentagons. Multiply by the remaining probability.

$= 0.15$

$P(\text{hexagon}) = \frac{7}{12} \times 0.36$ — Use the ratio of pentagons to hexagons to write the fraction for the number of hexagons. Multiply by the remaining probability.

$= 0.21$

LINK

Look back at Chapter 16 to remind yourself about ratios.

Outcomes and possibility spaces

The **possibility space** or **sample space** is a list or table showing all possible outcomes of an experiment or of combined experiments.

To show a possibility space:
- for one experiment, make a systematic list of all possible outcomes
- for two combined experiments, make a two-way table showing the combinations or end results.

Worked example

As part of a board game, players must select a card from a pack and roll an ordinary dice.

Half of the cards have a number 2 on them and the rest have a number 3.

The player multiplies the number on the card by the number on the dice to give their score.

1. Draw a table to show the possibility space.

    Dice

card	1	2	3	4	5	6
2	2	4	6	8	10	12
3	3	6	9	12	15	18

    Create a two-way table with the dice outcomes across the top, and card outcomes down the side.

    Multiply the numbers together to complete the possibility space.

2. Work out the probability of a player scoring more than 10.

    There are 12 equally likely outcomes.

    There are four outcomes which are more than 10.

    $$P(\text{more than 10}) = \frac{4}{12} = \frac{1}{3}$$

Relative frequency

If you don't know the probability of an event, you can do an experiment and use your results to **estimate** the probability. This way of estimating probability from an experiment is called finding the **relative frequency**.

$$\text{relative frequency} = \frac{\text{number of times the event happened}}{\text{total number of trials}}$$

Worked example

Jamie has an unfair dice.

He rolls the dice 60 times and scores a 5 on 20 rolls.

Estimate the probability of rolling 5 on Jamie's dice.

$$\text{estimate for } P(5) = \frac{\text{number of 5s Jamie rolled}}{\text{total number of rolls}}$$

$$= \frac{20}{60} = \frac{1}{3}$$

An unfair coin or spinner or dice is said to be **biased**.

Identify the event, and the total number of trials.

Substitute into the equation, and simplify the fraction.

> **REVISION TIP**
>
> Increasing the number of times an experiment is repeated results in a better estimate for the real probability. This is because, with more repeats, the relative frequency gets closer and closer to the real probability.

Knowledge

28 Probability

Expected results

The **expected frequency** of an event is the number of times you expect it to happen.

expected frequency = probability × number of trials

Worked example

1. Bishal rolls a dice 200 times and gets a 6 on 30 rolls.
   He claims this means the dice is biased.
   Use data to give a reason why he might be incorrect.

   With a fair dice, $P(6) = \frac{1}{6}$ — Identify the expected probability.

   expected frequency $= \frac{1}{6} \times 200 = 33.333...$ — Calculate the expected frequency.

   This is close to the actual result of 30, so he might be incorrect. The dice might be fair. — Compare the two and give a reason for your answer.

2. Ben rolls a dice six times and doesn't get a 6. He concludes that the dice is biased. Explain whether Ben is correct.
   Ben is wrong, because six rolls are not enough trials to draw a conclusion.

> **REVISION TIP**
> The expected frequency will not always match the actual outcome as it is just the most likely of many possible outcomes. If it is far off, though, the dice (or coin, spinner, etc.) is probably biased.

Tree diagrams

Tree diagrams can be used when considering probability involving two events.

- To find the probability of two events occurring together, **multiply** probabilities along the branches.
- To find the probability of one **or** the other happening, **add** the probabilities along the branches.

Worked example

In a board game, you pick a dice randomly from a bag and roll it.
The tree diagram shows the probabilities.

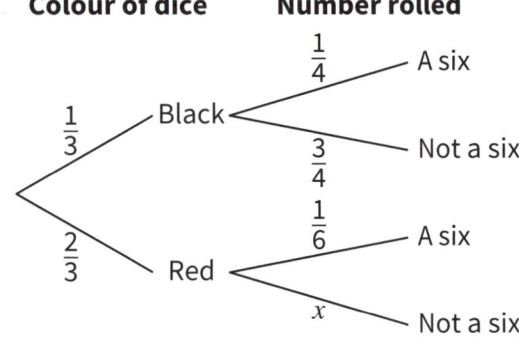

1. What is the value of x?
   $x = 1 - \frac{1}{6}$
   $= \frac{5}{6}$
   — Probabilities on each pair of branches must add up to 1.

2. Work out the probability of picking a black dice and rolling a 6.
   $P(\text{black and six}) = \frac{1}{3} \times \frac{1}{4} = \frac{1}{12}$
   — Multiply along the branches.

3. Work out the probability of rolling a 6.
   $P(\text{black and six}) = \frac{1}{12}$
   — Find the two possible ways to roll a 6.
   $P(\text{red and six}) = \frac{2}{3} \times \frac{1}{6} = \frac{2}{18}$
   $P(\text{six}) = \frac{1}{12} + \frac{2}{18} = \frac{7}{36}$
   — The two ways are mutually exclusive, so add them to find the probability of rolling a 6.

28

Conditional probability

Conditional probability is the probability of one event happening, given that another event has already occurred.

$$P(A \mid B) = \frac{P(A \cap B)}{P(B)}$$

The denominator is always the probability of the event you *know* has happened. In this case, B has occurred so the sample space is B. The numerator is the probability of the event you are interested in: A. The probability of this event happening in the sample space B is $P(A \cap B)$: the probability of both A and B happening.

> **REVISION TIP**
> When a question says 'without replacement', it will always involve conditional probability. A question about conditional probability often uses the word 'given'.

Worked example

60 people were asked how they travel to work.

- 33 of the people were men.
- 9 of the 23 people who travel by bus were women.
- 12 men walk to work.
- 19 people travel to work by car.

> **REVISION TIP**
> It is useful to write the information from a probability question in a two-way table.
>
> | | bus | walk | car | total |
> |---|---|---|---|---|
> | men | 14 | 12 | 7 | 33 |
> | women | 9 | 6 | 12 | 27 |
> | total | 23 | 18 | 19 | 60 |

1. Calculate the probability that they travel by car and are men

$$P(\text{car and men}) = \frac{7}{60}$$

Here the sample space is all 60 people.

2. Calculate the probability that they travel by car given they are men

$$P(\text{car and given men}) = \frac{7}{33}$$

Here the sample space is only the men.

Venn diagrams

$\mathcal{E}$ means the **universal set**, which contains all the numbers or objects.

$A \cap B$ means the **intersection** of sets A and B (numbers in **both** sets).

$A \cup B$ means the **union** of sets A and B (numbers in **either** set).

A' means the **complement** of set A (numbers **not** in set A).

$A \cap B$

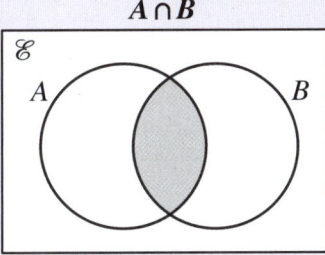

$A \cup B$

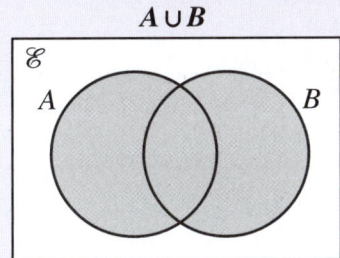

A'

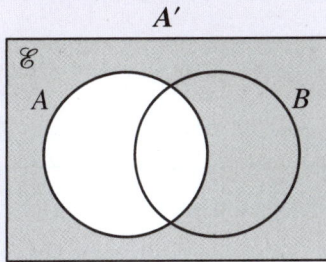

You can describe a set using curly brackets, for example, {1, 2, 3, 4} or {odd numbers}.

$x \in A$ means x is a member of the set A.

28 Knowledge 221

 # Knowledge

28 Probability

Venn diagrams

Worked example

There are 100 people at a swimming gala, 50 of whom are spectators.
There are three races: freestyle, backstroke and breaststroke.

10 people compete in all three races.
25 people compete in backstroke and breaststroke.
20 people compete in backstroke and freestyle.
12 people compete in freestyle and breaststroke.
30 people compete in freestyle.
1 person competes in only backstroke.

1. Work out the probability that a randomly selected person at the swimming gala will be competing in breaststroke.

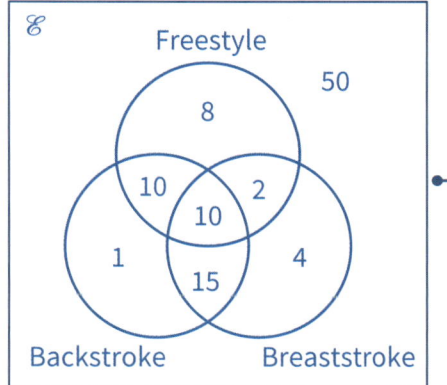

First draw a Venn diagram:
- identify the number of sets and draw three intersecting circles
- start in the middle with the numbers in all three sets
- complete the numbers in two sets
- complete the numbers in only one set
- complete the numbers in none of the sets.

$$P(\text{breaststroke}) = \frac{10 + 15 + 2 + 4}{100}$$

$$= \frac{31}{100}$$

Add the numbers in the four segments of the breaststroke set and divide by the total number of people at the event.

2. A person is selected from the backstroke racers. What is the probability they are also competing in freestyle?

$$P(\text{freestyle given backstroke}) = \frac{10 + 10}{10 + 10 + 15 + 1}$$

$$= \frac{20}{36} = \frac{5}{9}$$

Restrict the sample space to people competing in backstroke.

Key terms — Make sure you can write a definition for these key terms

biased   conditional probability   complement   estimate
expected frequency   intersection   mutually exclusive
possibility space (or sample space)   probability
probability scale   relative frequency   theoretical probability
tree diagram   union   universal set

28 Probability

Retrieval

28 Probability

Learn the answers to the questions below, then cover the answers column with a piece of paper and write as many as you can. Check and repeat.

#	Questions	Answers
1	What do we call a list or table showing all possible outcomes of an experiment or of combined experiments?	A possibility space or a sample space.
2	What is the estimated probability of an event happening called?	The relative frequency.
3	How do you calculate the relative frequency of an event?	Number of times the event happened divided by total number of trials.
4	How can you improve the accuracy of an estimated probability?	Increase the number of trials.
5	What is the expected frequency of an event?	The number of times you expect a event to happen.
6	How do you calculate probabilities along a branch of a tree diagram?	Multiply together the probabilities on the branch.
7	What do the probabilities of all possible mutually exclusive events add up to?	1
8	If one event affects the probability of another happening what do we call this?	Conditional probability.
9	What does $A \cup B$ mean?	A or B, the union of sets A and B.
10	What does $A \cap B$ mean?	A and B, the intersection of the sets A and B.
11	What does A' mean?	The complement of A, not A.

Previous questions

Now go back and use these questions to check your knowledge of previous topics.

#	Questions	Answers
1	What does HCF stand for?	Highest common factor.
2	What is ratio used for?	To compare two or more numbers.
3	When you square a negative number what type of number will your answer be?	Positive.
4	If $\tan \theta = 5$, how do you find θ?	$\theta = \tan^{-1} 5$
5	In the number 5^3, which value is the base?	5

Practice

Exam-style questions

28.1 A bag contains white, yellow, pink and orange counters. The table shows the probabilities of selecting each colour of counter from the bag.

Colour	White	Yellow	Pink	Orange
Probability	0.3	0.15	0.26	

(a) Work out the probability of not selecting a white counter or an orange counter. **[2 marks]**

(b) There are 200 counters in the bag. Work out how many orange counters there are. **[2 marks]**

28.2 A bag contains 4 red cubes, 6 yellow cubes and 5 blue cubes. Grace adds more red cubes to the bag until the probability of choosing a red at random is $\frac{1}{2}$.

How many red cubes does Grace add? **[2 marks]**

28.3 An unbiased 8-sided dice has the numbers 1, 2, 2, 3, 3, 4, 4 and 4 on its faces.

(a) If the dice is thrown once, what is the probability that it will land on 4? **[1 mark]**

(b) If the dice is thrown 40 times, how many times would you expect it to land on 3? **[2 marks]**

(c) The dice is thrown repeatedly and lands on the number 4 a total of 36 times. Estimate how many times the dice was thrown. **[2 marks]**

28.4 A manufacturer of light bulbs claims that 92% of light bulbs they produce last for longer than 25 000 hours. The light bulbs are sold in packs of two. A shop buys 500 packs.

Work out an estimate for the number of packs of light bulbs that will have exactly one light bulb that lasts longer than 25 000 hours. **[3 marks]**

> **EXAM TIP**
> It may help you to draw a tree diagram.

28.5 The frequency tree shows the outcomes of 40 people who took a driving test.

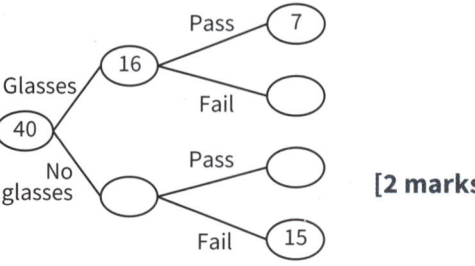

(a) Complete the frequency tree. **[2 marks]**

(b) Write the probability that a person chosen at random wears glasses and passes the test. **[1 mark]**

28.6 When dropped onto a hard surface, a plastic cup will land the right way up, on its side or upside down.

(a) Sheena is doing an experiment to find out which is the more likely outcome. She drops a plastic cup 50 times and records the results. Complete the table. **[2 marks]**

	Right way up	On its side	Upside down
Frequency	8		
Relative frequency		0.44	

(b) She then drops the plastic cup a further 100 times and it lands upside down 36 times.

Based on this, was the probability of the cup landing upside down higher in the first or second experiment? Show your working. **[2 marks]**

28.7 Nasim plays a game of darts with her friend. They then play a game of backgammon. The tree diagram below shows Nasim's probabilities of winning and losing each game.

The probability that Nasim wins both games is $\frac{1}{2}$.

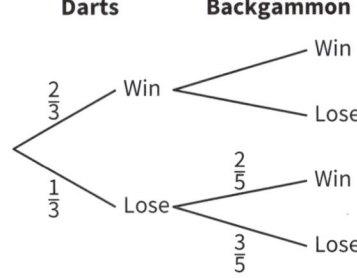

(a) Complete the two missing probabilities on the tree diagram. **[3 marks]**

(b) Work out the probability that Nasim's friend wins at most one of the two games. **[3 marks]**

28.8 Two people are chosen at random from a group of 4 French speakers and 7 English speakers. Using a tree diagram, show that the probability of selecting one French speaker and one English speaker is just over 50%. **[4 marks]**

28.9 At a school there are a total of 150 pupils in either year 9 or year 10. Each pupil studies one foreign language: French, German or Italian. Of the 87 pupils who study French, 35 are in year 9. 17 pupils in year 10 study German. A total of 12 pupils study Italian. There are 76 pupils in year 9.

One of the pupils is chosen at random. Work out the probability that the pupil is in year 9, given they study Italian. **[4 marks]**

Exam-style questions

28.10 A coin and a dice are repeatedly thrown together. Some of the results are shown in the two-way table.

Dice / Coin	Even number	Odd number
Heads		23
Tails	24	

The probability of getting tails, given that an odd number is rolled on the dice, is $\frac{26}{49}$. The probability of rolling an odd number on the dice, given that you get heads on the coin, is $\frac{23}{50}$. Complete the missing two values in the table. **[2 marks]**

> **EXAM TIP**
> Think about what the numerators and the denominators in the fractions represent.

28.11 The tree diagram shows the probabilities of a hockey player scoring a penalty on two consecutive attempts.

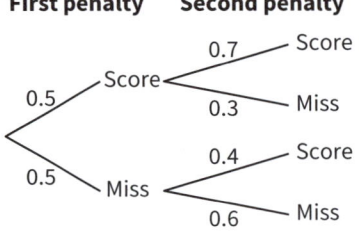

(a) Work out the probability that the hockey player scores with the 2nd penalty. **[2 marks]**

(b) Work out the probability that the hockey player scores with the 1st penalty given that they miss the 2nd penalty. **[3 marks]**

> **EXAM TIP**
> $P(A \text{ given } B) = \dfrac{P(A \text{ and } B)}{P(B)}$

28.12 Two events are scheduled to take place. The probability that events A and B both take place is 0.375. The probability that event A takes place, given that event B takes place, is 0.45.

Work out the probability that event B does not take place. **[4 marks]**

28.13 $\mathcal{E} = \{1, 2, 4, 8, 16, 25, 27, 64\}$, $A = \{\text{square numbers}\}$, $B = \{\text{cube numbers}\}$.

(a) Draw a Venn diagram for this information. **[4 marks]**

(b) A number x is chosen at random. Work out $P(x \in A \cap B)$. **[2 marks]**

28.14 $\mathcal{E} = \{1, 2, 3, 5, 6, 9, 10, 11, 17, 21, 25\}$
$F = \{\text{prime numbers}\}$, $G = \{\text{numbers greater than 10}\}$

(a) Draw a Venn diagram for this information. **[4 marks]**

(b) One of the numbers in the diagram is chosen at random. Find the probability that the number is in

(i) set $F \cap G$ (ii) set G' (iii) set F only. **[3 marks]**

28.15 The Venn diagram shows the number of customers who purchased apples, pears and bananas at a grocer's shop.

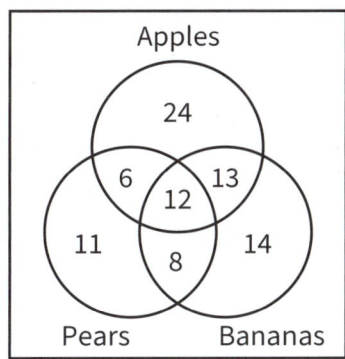

(a) Write the probability that a randomly chosen customer does not buy apples. [2 marks]

(b) Given that a randomly chosen customer bought bananas, what is the probability that they also bought apples? [2 marks]

EXAM TIP

'Given' means that the possibility space is restricted.

28.16 200 hundred people are asked which of the continents of Europe, North America and Africa they have visited during the last three years. 48 people had visited Africa. 5 had visited each of Europe, North America and Africa. Of the 58 people who had visited North America, 34 had not visited Europe or Africa. 16 people had visited both Europe and North America, whilst 19 people had visited both Europe and Africa. 22 people had visited none of the three continents.

(a) Work out the probability that a randomly chosen person visited Europe. [4 marks]

(b) Given that a randomly chosen person visited North America, work out the probability that they also visited Africa. [2 marks]

28.17 There are only green discs, yellow discs and pink discs in a bag. The ratio of the number of yellow discs to the number of pink discs is 7 : 11. Will takes a disc at random from the bag. The probability that the disc is green is 0.4.
Work out the probability that Will takes a pink disc. [3 marks]

28.18 There are only b black shirts and w white shirts on a clothes rail. A shirt is taken from the rail at random. The probability that the shirt is black is $\frac{4}{9}$. The shirt is then placed back on the rail. 4 more black shirts and 8 more white shirts are added to the rail. A shirt is taken from the rail at random. The probability that the shirt is black is $\frac{5}{12}$.
Find the number of black shirts and the number of white shirts that were on the rail originally. [5 marks]

Questions referring to previous content

28.19 (a) Vector $\mathbf{a} = \begin{pmatrix} 3 \\ -5 \end{pmatrix}$ and vector $\mathbf{b} = \begin{pmatrix} 9 \\ 4 \end{pmatrix}$
Calculate $3\mathbf{a} + 2\mathbf{b}$ [2 marks]

(b) Find vector $\mathbf{c}$ if $\mathbf{b} - 3\mathbf{c} = \mathbf{a}$ [3 marks]

28.20 Simplify fully $2\sqrt{80} + 3\sqrt{50} + 4\sqrt{45}$ [3 marks]

Knowledge

29 Tables, averages, and range

Averages

Average	Description	Example
Mean	mean = $\frac{\text{sum of all values}}{\text{number of values}}$ The mean is affected by outliers (unusually large or small values).	The mean of 3, 4, 8, 5 is: $\frac{3+4+8+5}{4} = \frac{20}{4}$ $= 5$
Median	The middle value when the data is put in order. If there are an even number of values, there will be two 'middle values'. The median is the mean of these two values. *Use the median instead of the mean if data has an outlier.*	The median of 3, 4, 5, 7 is: $\frac{4+5}{2} = \frac{9}{2}$ $= 4.5$
Mode	The most commonly occurring value. There can be no mode, one mode or multiple modes. *Use the mode if you have non-numerical data.*	The mode of 3, **4**, 7, 5, 3, **4**, 8, **4**, 6 is: 4
Range	The difference between the largest and smallest values.	The range of 3, 4, 7, 5 is: $7 - 3 = 4$

Ungrouped frequency tables

You can find the mean, median, mode and range when data is in an ungrouped frequency table.

Worked example

The table shows the number of sunny days per week over 21 weeks.

Number of sunny days per week	0	1	2	3	4	5	6	7	Total
Frequency	3	8	4	2	1	1	0	2	21
Number of days × frequency	0	8	8	6	4	5	0	14	45

1. Calculate the median.

   0, 0, 0, 1, 1, 1, 1, 1, 1, 1, ①, 2, 2, 2, 2, 3, 3, 4, 5, 7, 7

   Write out the data in an ordered list, and find the middle value(s).

2. Calculate the mode.

   The mode is 1.

   Find the number with the highest frequency in the table.

3. Calculate the mean.

   Mean = $\frac{45}{21}$

   $= 2.1$ (1 d.p.)

   Add a row to the frequency table for number of days × frequency.

   Then divide this total by the number of days to find the mean number of sunny days per week.

4. Calculate the range.

   Range = 7 − 0

   $= 7$

   highest value − lowest value

REVISION TIP
To find the mean, always divide by the sum of all the frequencies.

Grouped frequency tables

When data is given in a grouped frequency table, you can find:
- an estimate for the mean
- the **modal class** – the group with the highest frequency
- the **median class** – the group that the middle value lies in

For an ordered set of n data values, the median is the $\frac{1}{2}(n+1)$th data value.

Worked example

The lengths of 25 pieces of string are recorded in the table.

Length of string s (cm)	Frequency	Running total	Midpoint	Midpoint × frequency
$0 < s \leq 8$	12	12	4	48
$8 < s \leq 12$	6	18	10	60
$12 < s \leq 20$	7	25	16	112
Totals			–	220

1. Which is the modal class?

   $0 < s \leq 8$ is the modal class — Find the class with the highest frequency.

   — Add columns for the midpoint and midpoint × frequency, and calculate the values.

2. Which class does the median lie in? — Add a 'running total' column to the table.

   There are 25 pieces of string, so the median will be the 13th. — Identify the position of the median.

   13 lies between 12 and 18, so the 13th value lies in the $8 < s \leq 12$ class. — See which class the 13th value lies in.

3. Calculate an estimate for the mean.

   Estimate for mean $= \dfrac{220}{25}$ — Substitute the values into the equation for the mean.

   $= 8.8$

> **REVISION TIP**
>
> It is not possible to find the exact mean because we do not know all the exact values. So, estimate by assuming that, for each group, the length of each string is equal to the midpoint of the group.

# Knowledge

29 Tables, averages, and range

Median, quartiles, and interquartile range

For data in ascending order:

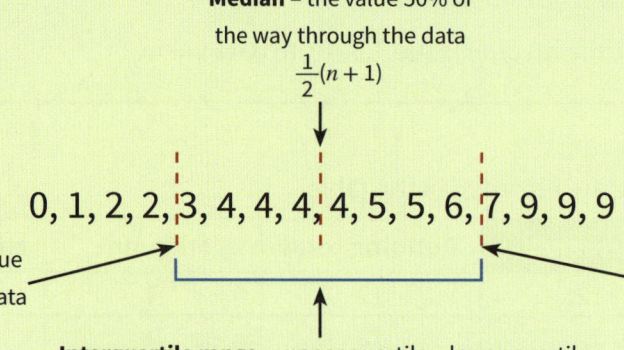

Median – the value 50% of the way through the data
$\frac{1}{2}(n+1)$

0, 1, 2, 2, 3, 4, 4, 4, 4, 5, 5, 6, 7, 9, 9, 9

Lower quartile (LQ) – the value 25% of the way through the data
$\frac{1}{4}(n+1)$

Upper quartile (UQ) – the value 75% of the way through the data
$\frac{3}{4}(n+1)$

Interquartile range = upper quartile − lower quartile

Worked example

Daisy played 11 games of football.
Here are the number of goals she scored in each game:

1 0 1 2 0 0 3 1 1 0 2

Find the interquartile range.

0, 0, 0, 0, 1, 1, 1, 1, 2, 2, 3 — Arrange the 11 items of data in ascending order.

$\frac{1}{4}(n+1) = 3 \Rightarrow$ 3rd value is 0 — Identify the position of the lower quartile.

$\frac{3}{4}(n+1) = 9 \Rightarrow$ 9th value is 2 — Identify the position of the upper quartile.

Interquartile range = 2 − 0
= 2 goals — Calculate the interquartile range.

Jessie also played 11 games of football.
Her median number of goals scored was 2 and her interquartile range was 3.
Compare Daisy's and Jessie's results.

$\frac{1}{2}(n+1) = 6 \Rightarrow$ 6th value is 1 — Calculate the position of the median.

Therefore, Daisy's median number of goals scored is 1.

Jessie's median is higher than Daisy's, — Compare the medians.
so Jessie scores more goals on average.

Daisy's interquartile range is smaller than Jessie's, — Compare the interquartile ranges.
so Daisy scores more consistently.

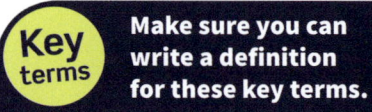 **Make sure you can write a definition for these key terms.**

interquartile range   lower quartile   mean   median
median class   modal class   mode   range   upper quartile

Retrieval

29 Tables, averages, and range

Learn the answers to the questions below, then cover the answers column with a piece of paper and write as many as you can. Check and repeat.

Questions | Answers

1. How would you find the mode from a frequency table? | Look for the interval or value with the highest frequency.
2. Do the groups need to be the same width in a grouped frequency table? | No.
3. What is the median of a set of data? | The middle value when the data is listed in order.
4. How do you calculate the range of a set of data? | Subtract the lowest value from the highest value.
5. How do you calculate the mean from a list of values? | Add up all the values and divide by the number of values.
6. What is the modal class of a grouped frequency table? | The class with the highest frequency.
7. What is the median class of a grouped frequency table? | The class that the middle value lies in.
8. What should you do if there are two middle values when finding the median? | Find the mean of the two middle values.
9. How do you calculate the mean of ungrouped data in a frequency table? | Multiply each value by its frequency, add up these values, and divide by the total of the frequencies.
10. What is an outlier? | A value which doesn't fit the trend.

Previous questions

Now go back and use these questions to check your knowledge of previous topics.

Questions | Answers

1. How do you represent an inequality on a number line where the value is NOT included? | Draw a hollow circle.
2. How do you find the gradient of a curve at a point? | Draw a tangent and find its gradient.
3. What is the hypotenuse? | The longest side of a right-angled triangle.
4. What is the general equation of a circle that has its centre at the origin? | $x^2 + y^2 = r^2$
5. What is the opposite of factorising? | Expanding.

Practice

Exam-style questions

29.1 The mean of 10 numbers is 63. The mean of 4 of the numbers is 51.

Work out the mean of the remaining 6 numbers. [3 marks]

29.2 71 people take part in a 5-km race.
The table shows their finishing times in minutes.

Time t (minutes)	Frequency
$20 < t \leq 25$	10
$25 < t \leq 30$	17
$30 < t \leq 35$	24
$35 < t \leq 40$	11
$40 < t \leq 45$	9

(a) Write the class interval that contains the median. [2 marks]

(b) Work out an estimate for the mean finishing time. Give your answer to the nearest minute. [3 marks]

(c) Explain why your answer to part **b** is only an estimate. [1 mark]

29.3 Jamie says, 'I can't estimate the mean score because I don't know the value of y.'

Score (x)	Frequency
$0 < x \leq 4$	$3y$
$4 < x \leq 8$	$7y$

Jamie is wrong. Work out an estimate of the mean score. [3 marks]

EXAM TIP
Insert extra columns in to the table.

29.4 The interquartile range of the numbers on these cards is 3.

3   5   6   7   8   9

What is the value of the missing number? [2 marks]

 29.5 Two basketball teams, A and B, each play 12 games.
Their scores are shown.

Team A: 87 85 82 76 76 100 95 63 123 78 90 99
Team B: 105 97 89 103 78 89 103 70 97 98 101 88

(a) Complete the table for Team A. [2 marks]

(b) Josh says, 'Team A have the highest score, so they must be the better team.' Is Josh correct? Explain your answer. [2 marks]

(c) Taisa says, 'Team A's scores are more varied than Team B's scores.' Is Taisa correct? Explain your answer. [2 marks]

Lowest Score	63.0
Lower Quartile	
Median	
Upper Quartile	
Highest Score	123.0

Here is the information for Team B.

Lowest Score	70.0
Lower Quartile	88.5
Median	97.0
Upper Quartile	102.0
Highest Score	105.0

Questions referring to previous content

29.6 Grass seed to cover an area of 3.66 m² costs £4.99.
Fabio needs grass seed for a lawn of 32 m².

How much will the grass seed cost Fabio?
Give your answer to the nearest pound. **[3 marks]**

 29.7 The sum of the interior angles in an irregular polygon is 720°

The smallest angle is 20° and the largest is 220°.

The angles form an arithmetic sequence. Find the other angles. **[3 marks]**

Knowledge

30 Charts and graphs

Bar charts

Bar charts are used for qualitative (categorical) data.

In a **bar chart**, the height of each bar shows the frequency. A **vertical line chart** uses lines instead of bars.

When drawing a bar chart
- label both axes
- make sure bars are the same width
- leave an equal gap between the bars.

Dual bar charts show two sets of data on the same bar chart.

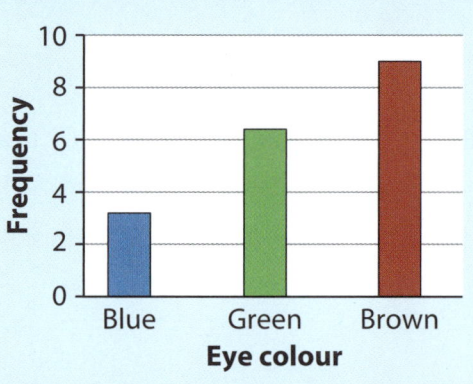

Worked example

The dual bar chart shows the average number of different drinks sold per day at a café. What is the most popular drink on weekdays?

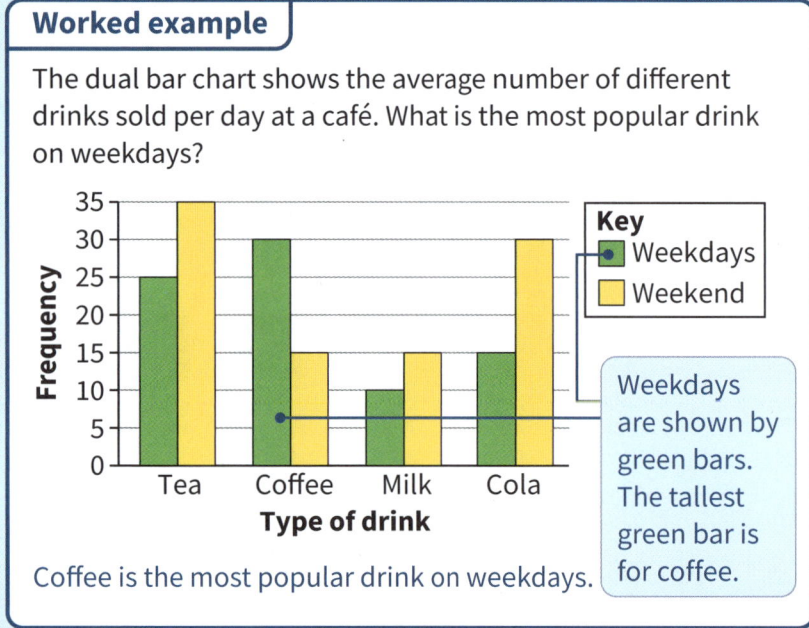

Coffee is the most popular drink on weekdays.

Weekdays are shown by green bars. The tallest green bar is for coffee.

The same data can be shown on a **composite bar chart**.

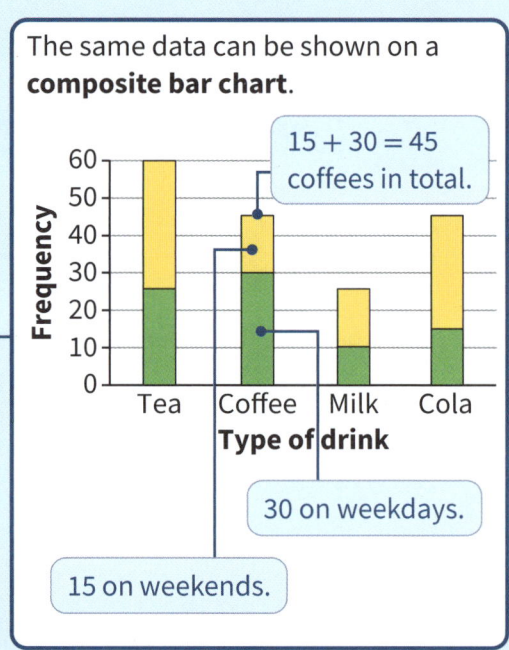

15 + 30 = 45 coffees in total.

30 on weekdays.

15 on weekends.

Pie charts

A **pie chart** shows categories as a proportion of the total.

The frequency or amount of something is shown as a sector of a circle.

The angle of the sector shows the proportion of that category in relation to the total.

angle of sector = proportion (as a fraction, decimal or percentage) × 360°

Colour of 25 cars parked on a street

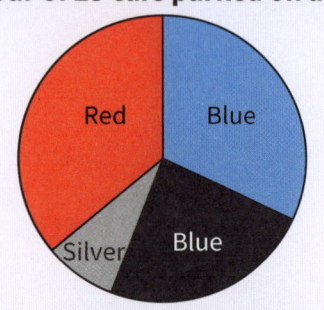

LINK

Refer back to Chapter 17 to remind yourself about proportions.

WATCH OUT

Be careful when comparing two pie charts with different totals!

30

Pie charts

Worked example

1. The vertical line chart shows the reasons for staff absence at a school.

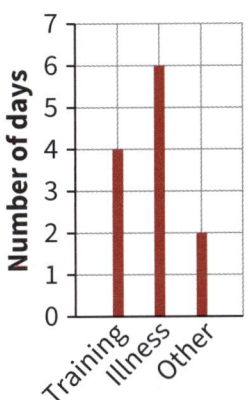

Reason for absence

Draw a pie chart to show this data.

Total = 4 + 6 + 2 = 12 ← Find the total number of days of absence.

Proportion = $\frac{4}{12} = \frac{1}{3}$ ← Find the proportion of absence days caused by training.

Angle of sector $\frac{1}{3} \times 360°$
= 120° ← The angles of all sectors in a circle always add up to 360°.

← Repeat to find the angles for 'Illness' and 'Other', and draw the pie chart.

Key
- Training
- Illness
- Other

← Remember to add a key.

2. A greengrocer has 81 pieces of fruit. He records the number of apples, bananas and pears.

He draws a pie chart to show the results. Calculate how many apples there were.

$\frac{a}{81} = \frac{160°}{360°} = \frac{4}{9}$

$a = \frac{4 \times 81}{9} = 36$ apples

Use equivalent fractions.

$\frac{\text{number of apples, } a}{\text{total pieces of fruit}} = \frac{\text{angle for apples}}{\text{total angle in circle}}$

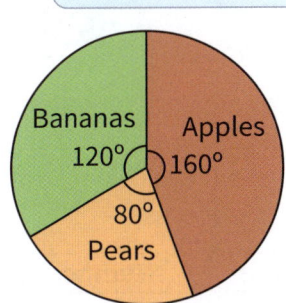

Frequency polygons

Frequency polygons represent grouped data.

The frequency of each class is plotted against its midpoint, then each point is joined with straight lines.

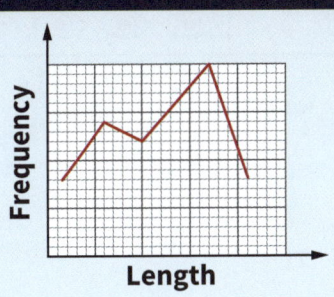

30 Knowledge **235**

Knowledge

30 Charts and graphs

Frequency polygons

Worked example

The grouped frequency table shows the heights of a sample of daisies.

Height of daisies (x cm)	Frequency
$2 < x \leq 6$	2
$6 < x \leq 10$	8
$10 < x \leq 14$	24
$14 < x \leq 18$	4

Draw a frequency polygon to represent the data.

Plot each frequency against the midpoint of each class. For example, the midpoint of the first class is 4 cm. Remember to join the points with straight lines.

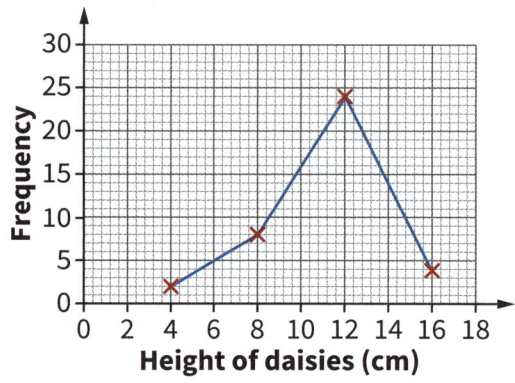

Pictograms

Pictograms are used for **qualitative data**. They use images to represent the data. For example, you could represent eye colour with a pictogram that uses pictures of eyes.

Key
👁 = 4 Children

Remember to add a key and use pictures that are simple to draw.

Line graphs

Line graphs (or time-series graphs) show how data changes over time. Time is always on the horizontal axis. For example, this graph shows the average temperature for each month for a year.

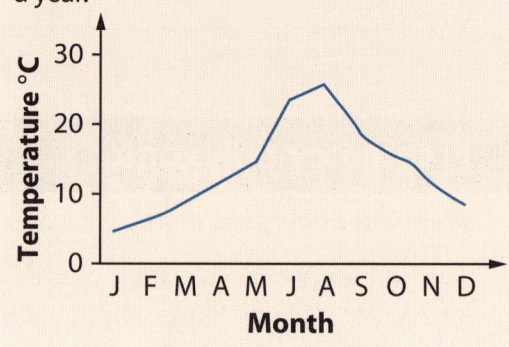

Key terms — Make sure you can write a definition for these key terms

bar chart   causation   correlation   dual bar chart   extrapolation
frequency polygon   interpolation   line graph   line of best fit
outliers   pictograms   pie chart   qualitative data
scatter graph   vertical line chart

Scatter graphs

Scatter graphs show the relationship between two sets of data (or variables), such as the age and height of a group of children.

In this example, each point would represent one child.

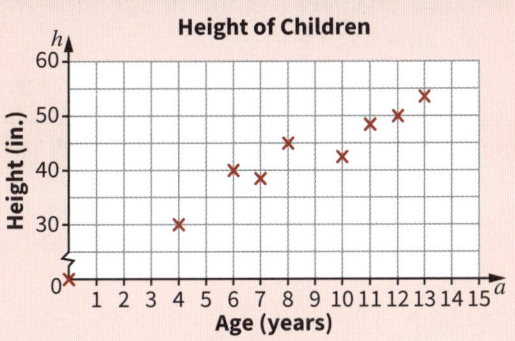

> **REVISION TIP**
>
> You can only plot quantitative data (numerical values) on a scatter graph. You could not plot hair colour against favourite food, for example, as there is no way that either quantity increases as you go along the axes.

Correlation describes how the data could be related. It can range from strong to weak, and be positive or negative:

Strong positive	Weak positive	No correlation	Weak negative	Strong negative
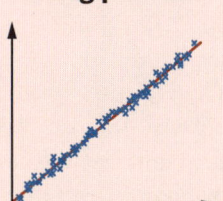	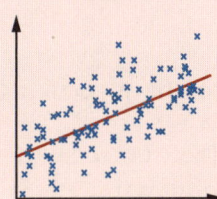	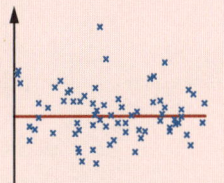	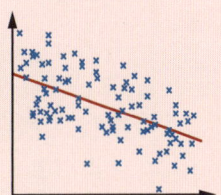	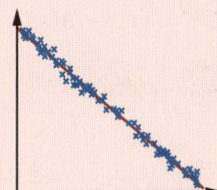

Causation means that one event causes another. Correlation between data sets does not necessarily mean that one event causes the other.

A **line of best fit** follows the trend of the data, ignoring **outliers**, and goes through (or near to) as many points as possible. The line of best fit should have roughly an equal number of data points on each side of it. It does not need to go through the origin.

Interpolation means using the line of best fit to estimate a value. It is a good method for making predictions within the line of best fit.

Extrapolation is predicting results outside the data (that is, outside your line of best fit).

> **REVISION TIP**
>
> Extrapolation can be useful but - be careful - it can also give unreliable results.

> **Worked example**
>
> The data shows the age of 10 chickens and the number of eggs each one laid over a month.
>
> How many eggs would you expect a 2-year-old hen to lay?
>
> | Age of the chicken | 1 | 2 | 2 | 3 | 3 | 4 | 4 | 4 | 5 | 5 |
> |---|---|---|---|---|---|---|---|---|---|---|
> | Number of eggs | 22 | 18 | 24 | 16 | 20 | 15 | 17 | 19 | 16 | 19 |
>
>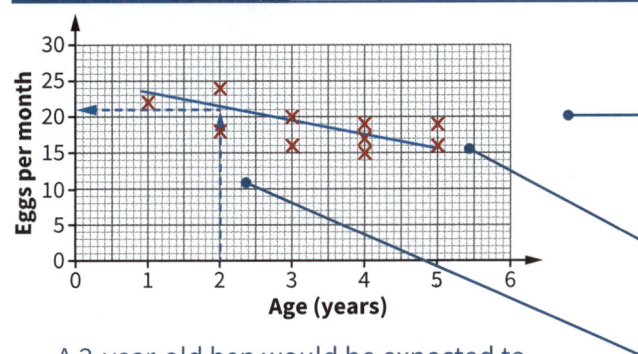
>
> Plot a scatter graph for the information given. Take care of the scale and make sure points are plotted accurately.
>
> Add a line of best fit by using your ruler to find a line through the middle of the points. Try to get the same number of points on each side. Ignore any outliers.
>
> Draw a line up from 2 on the x-axis to the line of best fit, then across to the y-axis.
>
> A 2-year-old hen would be expected to lay 21 eggs.

30 Knowledge 237

Retrieval

30 Charts and graphs

Learn the answers to the questions below, then cover the answers column with a piece of paper and write as many as you can. Check and repeat.

Questions / Answers

#	Question	Answer
1	What is correlation?	There is a relationship between two sets of data.
2	What is a line of best fit?	A line which follows the trend of the data.
3	What are the types of correlation?	Positive correlation and negative correlation.
4	What type of data can you represent using a pictogram?	Qualitative
5	What is extrapolation?	Making predictions outside the range of data.
6	Why do we need to be careful when extrapolating data?	It is not reliable because the relationship between the two variables might not hold for values outside the data.
7	What is the formula that is used to calculate the number of items in a pie chart sector?	number of items = (angle of sector ÷ 360°) × (total number of items)
8	When drawing a frequency polygon do you use the lower bound, upper bound, or midpoint of the class to plot the points?	The midpoint.

Previous questions

Now go back and use these questions to check your knowledge of previous topics.

Questions / Answers

#	Question	Answer
1	In algebra, what is a term? Give some examples.	A single number or variable, or numbers and variables multiplied together e.g. 2, x, $5y$ are all examples of terms.
2	Where is the line of symmetry on a quadratic graph?	On the vertical line that passes through the turning point.
3	When we divide two negative numbers what will be the result?	A positive number.
4	How do you convert cm^2 into mm^2?	Multiply by $10^2 = 100$.
5	What does magnitude mean?	Magnitude means size.

Practice 30

Exam-style questions

30.1 The table shows the monthly profits made by a shop over a period of months.

Month	Jan	Feb	Mar	Apr	May	Jun	Jul
Profit (£)	3000	3500	4200	4500	4000	5000	6000

(a) Construct a time series graph for this data. **[2 marks]**

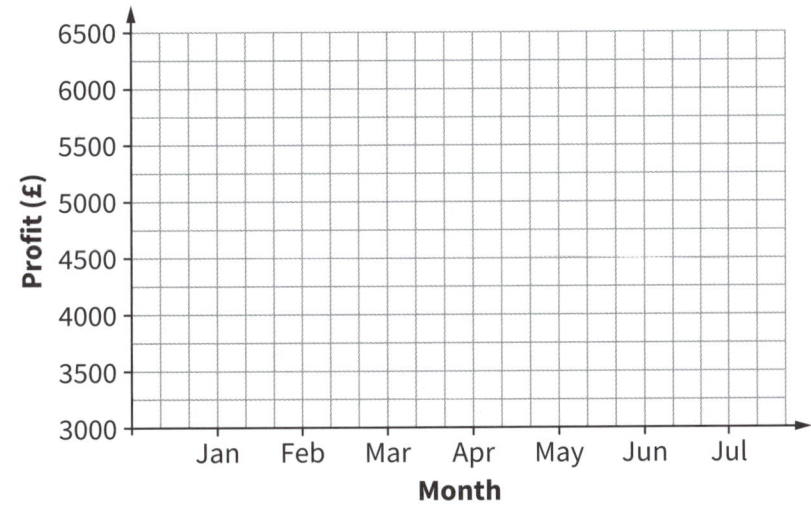

(b) Describe the general trend of the data. **[1 mark]**

(c) What might be misleading about this graph? **[1 mark]**

30.2 The time series graph shows the attendance figures at a football ground over an eight-week period.

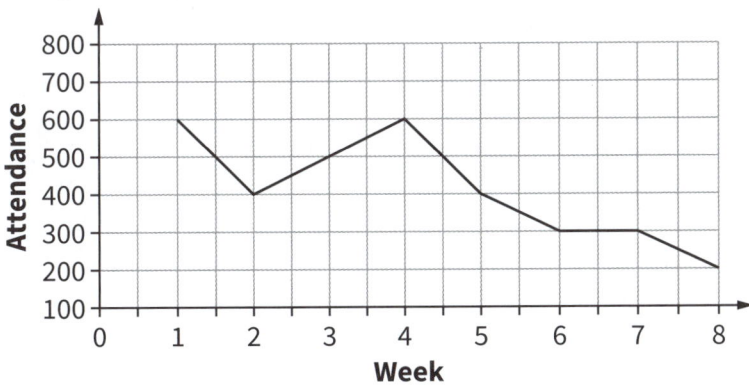

(a) Describe the general trend of the data. **[1 mark]**

(b) In which two weeks was attendance at its highest? **[1 mark]**

(c) Use the general trend to predict the attendance in Week 9. **[1 mark]**

Exam-style questions

30.3 The scatter graph shows the relationship between the uphill gradient of some hills and the speed of some cyclists.

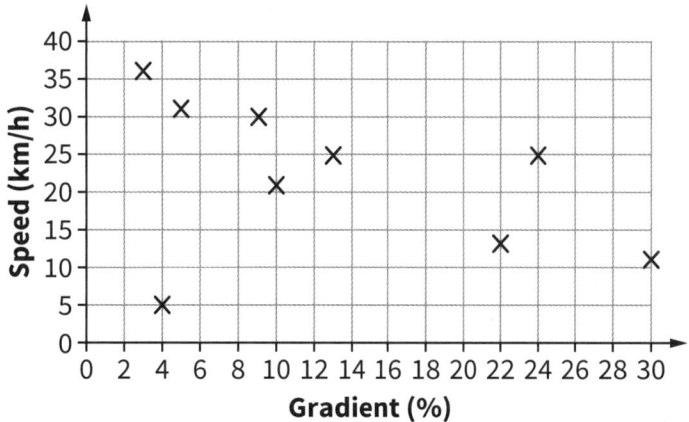

- **(a)** Give a reason why you might ignore an outlier when drawing a line of best fit. [1 mark]

- **(b)** What kind of correlation does the graph show? [1 mark]

- **(c)** Ignoring the outlier, draw a line of best fit on the scatter graph. [1 mark]

- **(d)** Use the line of best fit to estimate the uphill gradient of a hill if a cyclist is travelling at 15 km/h. [1 mark]

- **(e)** By how many km/h would you expect the speed of a cyclist to decrease for every 1% increase in the gradient of a hill? [2 marks]

- **(f)** Marta uses the line of best fit to estimate that a cyclist travelling up a hill with 40% gradient should be travelling at a speed of 5 km/h.

  Comment on Marta's assumption. [1 mark]

30.4 A scatter graph shows that there is positive correlation between the number of sandcastles made at a beach and the number of electric fans sold in a shop. Trevor says that the increase in the number of sandcastles causes an increase in the number of sales of electric fans.

Explain why Trevor is wrong. [1 mark]

30.5 Linda planted 400 flower bulbs. She planted a mix of daffodil, tulip and hyacinth bulbs. The incomplete table and pie chart show some information about these bulbs.

Type of bulb	Number planted
Daffodil	180
Tulip	
Hyacinth	
Total	400

Complete the table and pie chart. [3 marks]

30.6 The frequency polygon shows the masses of some puppies.

(a) Use the frequency polygon to complete the grouped frequency table. **[2 marks]**

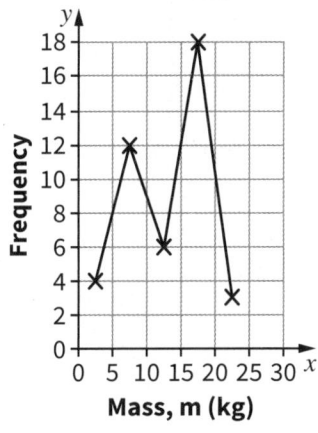

Mass m (kg)	Frequency
$0 \leq m <$	
$\leq m < 10$	12
$\leq m <$	6
$\leq m < 20$	
	3

(b) Estimate the mean mass of a puppy to the nearest kg. **[3 marks]**

30.7 Anastasia asks each of her 30 school friends how many of their school lessons they have enjoyed this week. The chart shows her results.

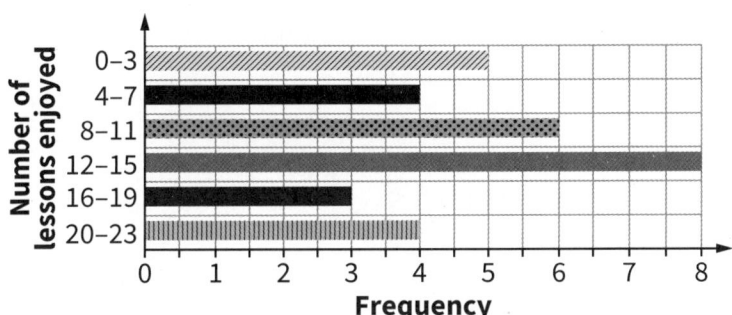

(a) What percentage of Anastasia's friends enjoyed more than 7 lessons? **[2 marks]**

(b) Estimate the mean number of lessons that Anastasia's friends enjoyed. **[4 marks]**

Number of lessons	Frequency
0–3	
4–7	
8–11	
12–15	
16–19	
20–23	

Question referring to previous content

30.8 Euler's polyhedral formula states that, in any polyhedron, the number of faces plus the number of vertices equals the number of edges plus 2.

Written mathematically: $F + V = E + 2$

A polyhedron has 10 vertices and 15 edges. Use the formula to find the number of faces and then draw an example of the 3D shape. **[3 marks]**

30 Practice

Knowledge

31 Data collection, cumulative frequency, and box plots

Types of data

	Definition	Example
Data	Any information you can collect.	The heights of people who live in London.
Population	The entire group of things that you can choose from.	All the people who live in London.
Sample	A smaller set within the population that you can collect data from.	One person from every street in London.
Biased sample	A sample that doesn't represent the population well.	A group of men from a basketball team.

In a **random sample**, each member of the population is equally likely to be chosen.

Data from a sample can be used to estimate properties of the whole population.

The larger the sample the more accurate the estimates.

> **Worked example**
>
> Jessica wants to know how many books the students in her school read per month.
>
> There are 1000 students in her school, so she wants to use a sample.
>
> She decides to go to the library and ask students how many books they have read that month.
>
> **(a)** Jessica asks 10 students in total, and finds that two of them have read more than five books that month.
> Use this data to estimate how many students in the entre school have read more than five books that month.
>
> $\frac{2}{10} = \frac{1}{5}$ ← Write the proportion of the sample who have read more than five books as a fraction, and simplify it.
>
> of the students have read more than five books.
>
> $\frac{1}{5} \times 1000$ ← Multiply the proportion in the sample by the number in the population to find the estimate.
>
> = 200 students
>
> **(b)** Give **two** reasons why your estimate for part **(a)** may not be accurate.
>
> Any two from:
> - The sample is not random (not all students have an equal chance of being chosen).
> - The sample may be biased (students in the library might be more likely to like reading than other students).
> - The sample size of 10 students is too small to be representative of the whole school.

Capture-recapture

The **capture-recapture** method is a way to estimate the size of a population of animals.

When using the capture-recapture method:
- it is important to allow time between samples for the marked and unmarked animals to mix (but not too long as you don't want to have to consider the birth rate or death rate)
- you need to be confident that the animals can't escape the environment.

31

Capture-recapture

To estimate the size of an animal population (N):

$$N = \frac{nK}{k}$$

Where:
K = animals in the first sample that are marked
k = marked animals found in the second sample
n = size of the second sample.

Worked example

Elijah is trying to estimate the population of rabbits in a fenced area.

He captures a sample of five rabbits, marks each of them and then releases them.

The next day he recaptures a second sample of eight rabbits and finds two of them are marked.

1. Estimate the number of rabbits in the area.

$$N = \frac{nK}{k}$$
$K = 5, k = 2$ and $n = 8$, so
$$N = \frac{8 \times 5}{2}$$
$$= 20$$

Identify the information you have and substitute the values into the equation to estimate animal population size.

2. What assumptions have you made?
   - the samples are random
   - the marked and unmarked rabbits mix up
   - the population hasn't changed overnight

Three possible assumptions are listed, but you only need to give two.

REVISION TIP

You don't have to remember the formula – you can assume that the proportion of marked animals in the recaptured sample is the same as the proportion in the whole population. Here, the proportion of marked rabbits is $\frac{2}{8}$. Assuming that $\frac{2}{8}$ of all the rabbits in the area have been marked, the total number of rabbits in the area is $5 \times 4 = 20$.

Cumulative frequency graphs

Cumulative frequency graphs show the total frequency up to each point.

Worked example

The table shows the heights of some horses.

Height, h (m)	Frequency	Cumulative frequency
$2.1 < h \leq 2.2$	2	2
$2.2 < h \leq 2.3$	3	5
$2.3 < h \leq 2.4$	7	12
$2.4 < h \leq 2.5$	7	19
$2.5 < h \leq 2.6$	1	20

(a) Draw a cumulative frequency graph for this information.

First, add a cumulative frequency column to the table.

31 Knowledge

Knowledge

31 Data collection, cumulative frequency, and box plots

Cumulative frequency graphs

Worked example

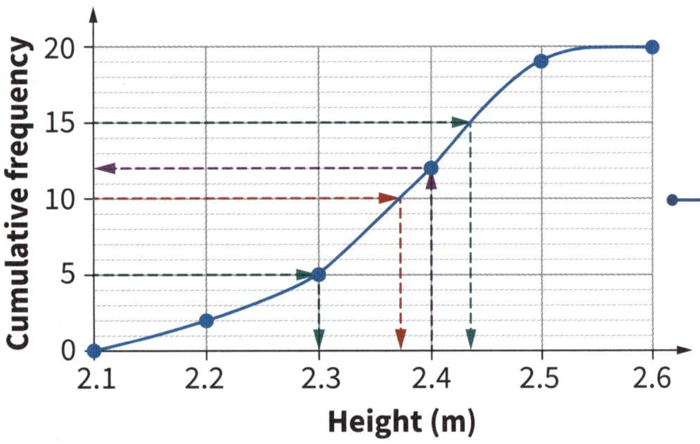

Plot a graph with Cumulative frequency on the y-axis, and the end-point of each height class on the x-axis. Join the points with a smooth curve.

(b) Estimate the median height.

Median ≈ 2.37 m

There are 20 horses in total, so for the median, draw a line from 10 on the y-axis, across to the graph line, then down to the x-axis.

(c) Estimate the interquartile range in heights.

$LQ ≈ 2.30\ m : UQ ≈ 2.44\ m$

Interquartile range = 2.44 − 2.30

= 0.14 m

For the LQ and UQ, draw lines from 5 and 15 on the y-axis across to the graph, and read the value on the x-axis.

(d) Estimate the percentage of horses measuring less than 2.4 m.

12 horses, which is 60% of 20

Draw a line up from 2.4 m on the x-axis.

Then calculate the number of horses as a percentage of the total number of horses.

Box plots

A **box plot** can be used to show the median, range and interquartile range of data and is useful for comparing sets of data.

To identify the quartile values:

Lower quartile = $\frac{1}{4}$ (number of values in set + 1)

Upper quartile = $\frac{3}{4}$ (number of values + 1)

Features of a box plot:

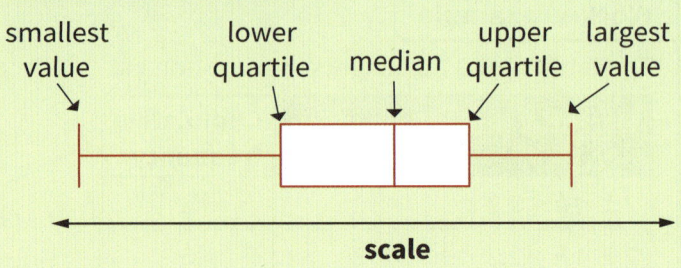

Key terms — Make sure you can write a definition for these key terms

biased sample box plot cumulative frequency graph
frequency density histogram capture-recapture
population random sample

Box plots

Worked example

The heights of a group of children are (in cm):

144 125 141 135 136 132 142 160 133

(a) Identify any outliers in the data.

An outlier is any value that more than $1.5 \times IQR$ above the upper quartile or $1.5 \times IQR$ below the lower quartile.

125, 132, 133, 135, 136, 141, 142, 144, 160 — *List the heights in order from smallest to largest.*

There are 9 children:

Median = 5th value = 136 cm — *Identify the median.*

Lower quartile:
$\frac{1}{4}((9+1)) = 2.5$
mean of 2nd and 3rd values = 132.5 cm

Upper quartile
$\frac{3}{4}(9+1) = 7.5$
mean of 7th and 8th values = 143 cm

— *Find the quartiles.*

$IQR = 143 - 132.5 = 10.5$ cm — *Calculate the IQR*

$1.5 \times IQR = 1.5 \times 10.5$
$= 15.75$ — *Calculate $1.5 \times IQR$*

Lower quartile $- 1.5 \times IQR = 132.5 - 15.75$
$= 116.75$

Upper quartile $+ 1.5 \times IQR = 143 + 15.75$
$= 158.75$

— *Calculate the values below and above which any data would be called outliers.*

125, 132, 133, 135, 136, 141, 142, 144, (160)

— *160 cm lies above 158.75 cm so is an outlier.*

(b) Represent this data using a box plot.

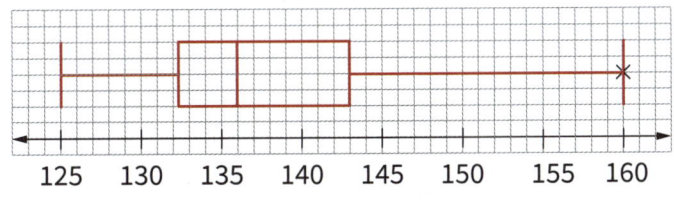

— *Use the values to draw the box plot.*

Histograms

Histograms can be used to represent grouped data. They are particularly useful when the class widths are different.

In a histogram:

- the *y*-axis shows the **frequency density**

  frequency density $= \dfrac{\text{frequency}}{\text{class width}}$

- the **area** of a bar is proportional to the frequency of the class.

Knowledge

31 Data collection, cumulative frequency, and box plots

Histograms

Worked example

The table shows the results of a survey of the heights of 100 Year 10 students.

Height, x (cm)	Frequency	Frequency density
$100 < x \leq 110$	13	1.3
$110 < x \leq 135$	24	0.96
$135 < x \leq 150$	36	2.4
$150 < x \leq 160$	17	1.7
$160 < x \leq 200$	10	0.25

(a) Draw a histogram of this data.

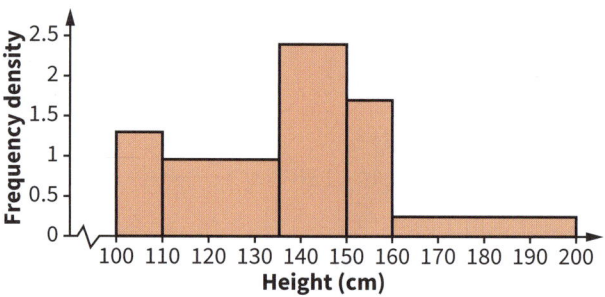

Start by calculating the frequency density.

(b) Find an estimate of the median to 1 d.p.

Median value = 50th value — Median value = middle of the data set (100 ÷ 2).

Median falls in the interval $135 < x \leq 150$ — Find the interval the median falls in by adding the frequencies.

Median value = 50 − 13 − 24 — Find the position of the median in this interval by subtracting the frequencies of the previous intervals.

= 13th value in the interval

$= \frac{13}{36} \times 15 + 135$ — Find the 13th value using the interval width of 15. Assume that the group is equally spaced, so all 36 values are the same distance from each other.

= 140.4

Worked example

The histogram shows the times that some children took to complete a short test.

Three children completed the test in under 6 seconds.

Work out the number of children that took between 6 and 10 seconds to complete the test.

3 children completed the test in under 6 s, so for the 0-6 s class,

frequency density (f.d.) = $\frac{\text{frequency (f)}}{\text{class width (c.w.)}} = \frac{3}{6}$

= 0.5

So the frequency density axis has a scale of 1 square = 0.5

Hence the number of children that took between 6 and 10 seconds = 2.5 × 4 = 10 children

Retrieval 31

31 Data collection, cumulative frequency, box plots

Learn the answers to the questions below, then cover the answers column with a piece of paper and write as many as you can. Check and repeat.

Questions | Answers

1. Why should you not use a biased sample? | It will not represent the population well.

2. Why do we use the capture–recapture method? | To estimate the size of a population.

3. What is a cumulative frequency graph? | A graph that shows the total frequency up to each point and can be used to estimate the median and quartiles.

4. What do you plot on the x-axis of a cumulative frequency graph from grouped data? | The end point of each class interval.

5. How do you estimate the median? | Draw a line across from the y-axis at 50% of the total frequency, and then down to the x-axis. This value is the median.

6. How do you estimate the upper and lower quartiles from a cumulative frequency graph? | Draw lines across from the y-axis at 25% and 75% of the total frequency, and then down to the x-axis.

7. What is a box plot used for? | For comparing sets of data.

8. What type of data is displayed on a histogram? | Grouped data.

Previous questions

Now go back and use these questions to check your knowledge of previous topics.

Questions | Answers

1. If the ratio $a:b$ is $10:7$, what is the ratio of $b:a$? | $7:10$

2. What are the conditions for a triangle to be congruent? | SSS, SAS, ASA, RHS

3. What is the cosine rule to find an angle? | $\cos A = \dfrac{b^2 + c^2 - a^2}{2bc}$

4. What is the exact value of: $\sin 60°$? | $\dfrac{\sqrt{3}}{2}$

5. What is the formula used to find the area of a triangle using trigonometry? | $A = \dfrac{1}{2} ab \sin C$

31 Retrieval 247

Practice

Exam-style questions

31.1 Sam wants to predict the football results for upcoming matches. She uses a list of results for matches that took place five years ago.

Explain why the data might not be reliable. **[1 mark]**

31.2 180 sixth form students want to celebrate their end-of-course exam results. A sample of 40 students is taken to decide how they should celebrate. Their preferences are shown in the table.

Type of celebration	Frequency
Cinema trip	11
Party	10
Meal out	11
Day at funfair	8

(a) Work out how many of the 180 students you think would like a day at the funfair. **[2 marks]**

(b) State one assumption that you have made in your answer to part **a**. **[1 mark]**

31.3 (a) Write what is meant by a random sample. **[1 mark]**

Harriet wants to do a questionnaire about the favourite foods of students in her school. She selects five of her best friends to take part.

(b) Identify
　(i) the population **[1 mark]**
　(ii) the sample. **[1 mark]**

(c) Give a reason why her sample might be biased. **[1 mark]**

(d) Write two ways that her sample can be improved. **[2 marks]**

31.4 A team of researchers want to know the approximate number of gulls at a seaside resort. On Tuesday they catch 50 gulls. They put a tag on each gull and let it go.

On Wednesday they catch 234 gulls. They find that 18 of these gulls have a tag.

(a) Work out an estimate for the number of gulls at the resort. **[3 marks]**

(b) State one assumption you have made in your answer to part **a**. **[1 mark]**

One of the researchers suspects that the tags on some of the gulls fell off after they were tagged.

(c) Describe what effect this would have on your answer to part **a**. Give a reason for your answer. **[1 mark]**

31.5 The lengths of 25 aardvarks, in metres, are shown below.

1.3 1.6 1.8 1.9 1.1 1.4 1.5 1.0 0.8 0.8 1.5 1.7
1.7 1.9 1.4 1.1 0.8 2.0 1.7 2.0 1.3 1.4 1.8 0.9 1.1

Draw a box plot to represent this information. **[4 marks]**

EXAM TIP
Remember to start with an ordered list.

31.6 The table shows some information about the time taken in minutes for a year 10 class to complete their maths homework.

Minimum	Interquartile range	Maximum	Lower quartile	Median
22	8	48	32	34

(a) Draw a box plot to represent this information. **[2 marks]**

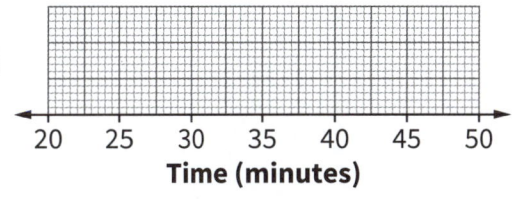

The box plot shows the distribution of times for a year 11 class to complete their homework.

(b) Compare the distribution of the year 11 times to the year 10 times. **[3 marks]**

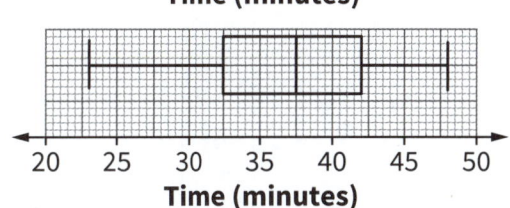

31.7 The frequency table shows the times taken for 200 students to get ready for school.

(a) On the grid, draw a cumulative frequency graph to represent this information. **[2 marks]**

Time (t minutes)	Frequency
$0 < t \leq 5$	30
$5 < t \leq 10$	74
$10 < t \leq 15$	56
$15 < t \leq 20$	22
$20 < t \leq 25$	18

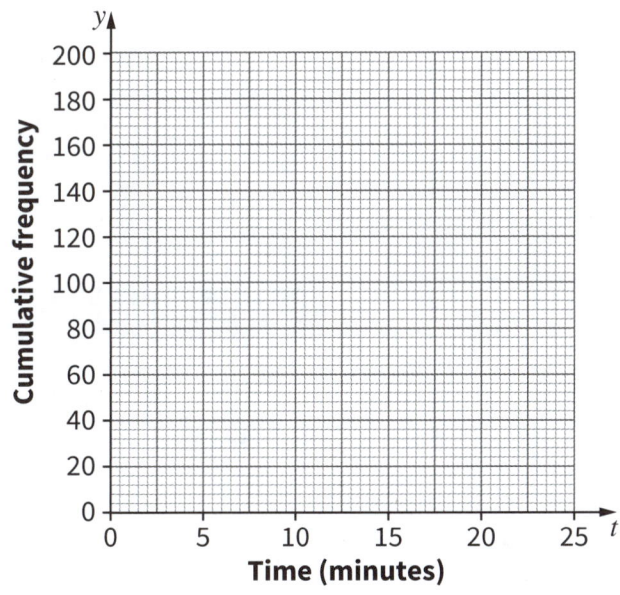

(b) Find the median time. **[1 mark]**

(c) Neve says, 'More than 30% of pupils take less than 8 minutes to get ready for school.'

Is Neve correct? You must show how you got your answer. **[3 marks]**

Exam-style questions

31.8 The frequency table shows the speeds, in km/h, of some cars on a motorway.

(a) Complete the missing information in the table and in the histogram. **[4 marks]**

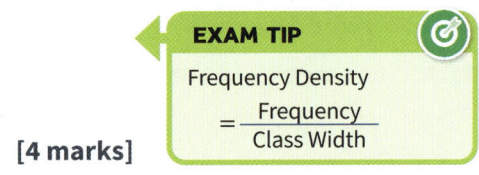

EXAM TIP

Frequency Density = Frequency / Class Width

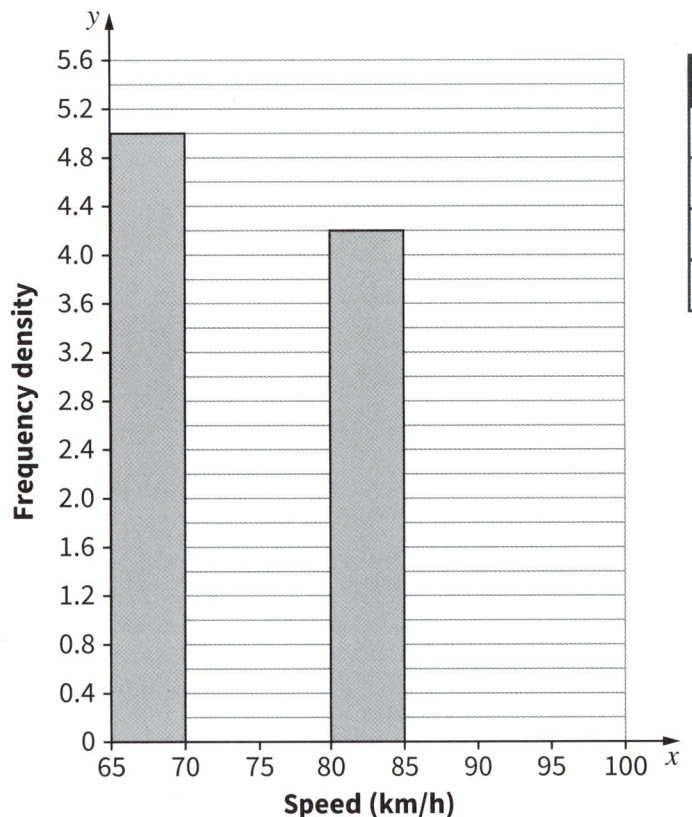

Speed (s km/h)	Frequency
$65 < s \leq 70$	
$70 < s \leq 80$	44
$80 < s \leq 85$	
$85 < s \leq 100$	18

(b) Estimate the mean speed. **[3 marks]**

(c) Estimate the median speed, stating any assumptions you have made. **[3 marks]**

EXAM TIP

Find the class that contains the median, and then use a proportional method.

31.9 The table shows information about the masses, in kg, of some dogs.

(a) Nanveet says, 'The mean might not be the best average to use for this data.'

Give a reason to support Nanveet's claim. **[1 mark]**

(b) Find the class interval that contains the median. **[1 mark]**

(c) On the grid, on the next page draw a frequency polygon for the information in the table. **[2 marks]**

Mass (m kg)	Frequency
$0 < m \leq 4$	3
$4 < m \leq 8$	6
$8 < m \leq 12$	5
$12 < m \leq 16$	8
$16 < m \leq 20$	20

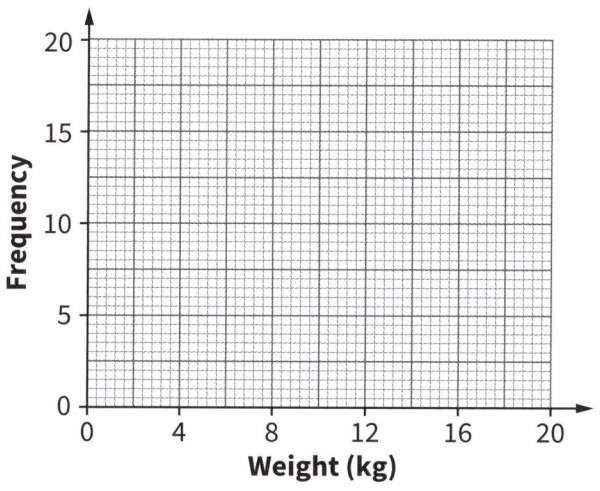

31.10 The cumulative frequency graph shows information about Test 1 scores for 32 pupils.

The lowest score was 10 and the highest score was 48.

(a) On the grid below, draw a box plot to represent this information. **[3 marks]**

The box plot shows the information about the same pupils' scores for Test 2.

(b) Cain says, 'The median score for the second test is higher, so that test must have been easier.'

Is Cain right? You must give a reason for your answer. **[2 marks]**

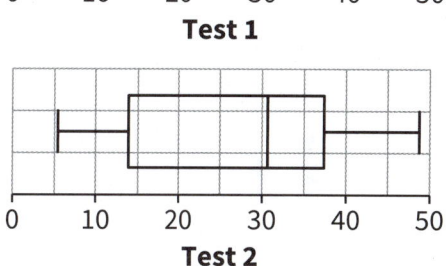

Question referring to previous content

31.11 Each of these equations matches to one of the graphs.

$y = \dfrac{1}{x}$ $y = x^2$ $y = x^3$ $y = x$

Write the correct equation below each graph. **[2 marks]**

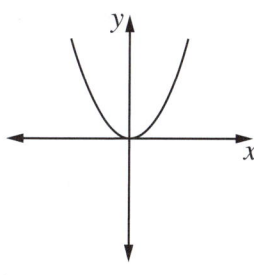   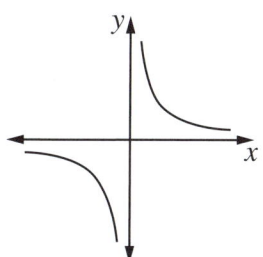

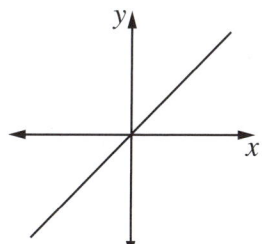   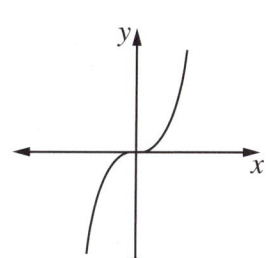

Great Clarendon Street, Oxford, OX2 6DP, United Kingdom

Oxford University Press is a department of the University of Oxford. It furthers the University's objective of excellence in research, scholarship, and education by publishing worldwide. Oxford is a registered trade mark of Oxford University Press in the UK and in certain other countries.

© Oxford University Press 2023

Written by Naomi Bartholomew-Millar, Paul Hunt and Victoria Trumper

The moral rights of the authors have been asserted

The publisher would also like to thank Katie Wood and Jemma Sherwood for their work on the first edition of Edexcel GCSE Maths Higher Revision Guide (978-138-200651-4) and Revision Workbook (978-138-200652-1) on which this revision guide is based.

First published in 2023

All rights reserved. No part of this publication may be reproduced, stored in a retrieval system, or transmitted, in any form or by any means, without the prior permission in writing of Oxford University Press, or as expressly permitted by law, by licence or under terms agreed with the appropriate reprographics rights organization. Enquiries concerning reproduction outside the scope of the above should be sent to the Rights Department, Oxford University Press, at the address above.

You must not circulate this work in any other form and you must impose this same condition on any acquirer

British Library Cataloguing in Publication Data
Data available

978-138-203985-7

978-1-382-06901-4 (ebook)

10 9 8 7 6 5 4 3 2

The manufacturing process conforms to the environmental regulations of the country of origin.

Printed in China by Shanghai Offset Printing Products Ltd

Acknowledgements
Artworks: QBS Learning

The publisher would also like to thank Deb Friis and Katherine Pate for sharing their expertise and feedback in the development of this resource.

Although we have made every effort to trace and contact all copyright holders before publication this has not been possible in all cases. If notified, the publisher will rectify any errors or omissions at the earliest opportunity.

Links to third party websites are provided by Oxford in good faith and for information only. Oxford disclaims any responsibility for the materials contained in any third party website referenced in this work.

The manufacturer's authorised representative in the EU for product safety is Oxford University Press España S.A. of El Parque Empresarial San Fernando de Henares, Avenida de Castilla, 2 – 28830 Madrid (www.oup.es/en or product.safety@oup.com).OUP España S.A. also acts as importer into Spain of products made by the manufacturer.